AF768446

Silicon

Editor: J. Grabmaier

With Contributions by
T. F. Ciszek J. Dietl W. Dietze
D. Helmreich W. Keller A. Mühlbauer
E. Sirtl F. V. Wald

Springer-Verlag
Berlin Heidelberg GmbH

Managing Editor

Prof. Dr. H. C. Freyhardt, Kristall-Labor der Physikalischen Institute, Lotzestr. 16–18, D-3400 Göttingen
and Institut für Metallphysik der Universität Göttingen, Hospitalstr. 12, D-3400 Göttingen

Editorial Board

Prof. *T. Arizumi,* Department of Electronics, Nagoya University, Furo-cho Chikusa-Ku, Nagoya 464, Japan
Dr. *W. Bardsley,* Royal Radar Establishment, Great Malvern, England
Prof. *H. Bethge,* Institut für Festkörperphysik und Elektronenmikroskopie, Weinberg, 4010 Halle/Saale, DDR
Prof. *A. A. Chernov,* Institute of Cristallography, Academy of Sciences, Leninsky Prospekt 59, Moscow B – 117333, USSR
Dr. *J. Grabmaier,* Siemens AG, Forschungslaboratorien, Postfach 801709, 8000 München 83, Germany
Prof. *S. Haussühl,* Institut für Kristallographie der Universität Köln, Zülpicherstr. 49, 5000 Köln, Germany
Prof. *R. Hoppe,* Institut für Anorganische und Analytische Chemie der Justus-Liebig-Universität, Heinrich-Buff-Ring 58, 6300 Gießen, Germany
Prof. *R. Kern,* Université Aix-Marseille III, Faculté des Sciences de St. Jérome, 13397 Marseille Cedex 4, France
Dr. *R. A. Laudise,* Bell Laboratories, Murray Hill, NJ 07974, U.S.A.
Prof. *R. Nitsche,* Kristallographisches Institut der Universität Freiburg, Hebelstraße 25, 7800 Freiburg, Germany
Prof. *A. Rabenau,* Max-Planck-Institut für Festkörperforschung, Heisenbergstr. 1, 7000 Stuttgart 80, Germany
Dr. *C. J. M. Rooijmans,* N. V. Philips' Gloeilampenfabrieken Research Laboratories, 5621 CP Eindhoven, The Netherlands
Prof. *W. B. White,* Materials Research Laboratory, The Pennsylvania State University, University Park, PA 16802, U.S.A.
Prof. *A. F. Witt,* Massachusetts Institute of Technology, Cambridge, MA 02139, U.S.A.
Dr. *F. W. Young,* Jr., Solid State Division, Oak Ridge National Laboratory, P.O. BOX X, Oak Ridge, TN 37830. U.S.A.

ISBN 978-3-642-68177-6 ISBN 978-3-642-68175-2 (eBook)
DOI 10.1007/978-3-642-68175-2

Library of Congress Cataloging in Publication Data. Main entry under title: Silicon/with contributions by T. F. Ciszek. (Crystals-growth, properties, and applications; 5) Bibliography: p. Includes index. 1. Silicon crystals-Addresses, essays, lectures. I. Ciszek, T. F. (Ted F.) II. Series. QD921.S5392 661'.0683 81-9112 AACR2

This work is subject to copyright. All rights are reserved, whether the whole or part of materials is concerned, specifically those of translation, reprinting, re-use of illustrations, broadcasting, reproduction by photocopying machine or similar means, and storage in data banks. Under § 54 of the German Copyright Law where copies are made for other than private use, a fee is payable to „Verwertungsgesellschaft Wort", Munich.

© by Springer-Verlag Berlin Heidelberg 1981
Originally published by Springer-Verlag Berlin Heidelberg New York 1981
Softcover reprint of the hardcover 1st edition 1981

The use of registered names, trademarks, etc. in this publication does not imply, even in the absence of a specific statement, that such names are exempt from the relevant protective laws and regulations and therefore free for general use.

2152/3140-543210

5 Crystals

Growth, Properties, and Applications

Managing Editor: H. C. Freyhardt

Editors: T. Arizumi, W. Bardsley, H. Bethge
A. A. Chernov, H. C. Freyhardt, J. Grabmaier
S. Haussühl, R. Hoppe, R. Kern, R. A. Laudise
R. Nitsche, A. Rabenau, C. J. M. Rooijmans
W. B. White, A. F. Witt, F. W. Young, Jr.

Table of Contents

Float-Zone Grown Silicon

Wolfgang Dietze[1], Wolfgang Keller[1] and Alfred Mühlbauer[2]

[1] Siemens AG, Frankfurter Ring 152, D-8000 München 46
[2] Institut für Elektrowärme, Universität Hannover, Wilhelm-Busch-Str. 4, D-3000 Hannover

Float-zone grown silicon today is a material of unsurpassed crystal quality and purity, which is of paramount importance as starting material for the electronics industry. This article describes the historical development up to the present state, gives a survey of the processes leading to the starting material, the polycrystalline silicon rods, and depicts the usual technology of the float-zone process, especially the dislocation-free crystal pulling technique. Moreover, silicon crystallography, the properties of silicon crystals, and the crystal defects, which are found in float-zone grown silicon, are described. Of course, important doping methods for preparing semiconductor silicon are referred to. Finally, a comparison is made with the other important crystal growth technique, crucible pulling, and the present and future trends are outlined. A discussion of the method is given in[1)].

Crystals, Growth, Properties and Applications, Vol. 5
© Springer-Verlag Berlin Heidelberg 1981

1 Historical Survey

Prior to the industrial application of the float-zone technique in the second half of the fifties[1)], crystals of commercial dimensions were grown by the Czochralski method[2)], i.e. the melt was contained in a crucible and the crystal pulled out of this crucible. According to this method, semiconductor crystals were grown starting about 1950, at first germanium, but soon also silicon[3)]. Today, still about 80% of semiconductor single crystals are produced by the Czochralski method. Because of the contaminating crucible walls, however, the resulting crystals can only have a limited purity which cannot be exceeded.

In the early fifties, Pfann[4] started his zone-melting work which led to new ways of producing pure crystals. His method made it possible to grow crystals of sufficient purity for semiconductor devices, because of the repeatability of the process. In the early times of semiconductor work, the quality of the starting material was not sufficient for device purposes. Hence, multiple zone-melting purification was essential, e.g. for the production of germanium starting material. Based on zone refining, Theuerer[5)] invented the crucible-free floating-zone method excluding contamination from crucible walls.

Independently, Keck and Golay[6)] and Emeis[7)] grew silicon crystals by the floating-zone method. Müller[8)] and Keck and coworkers[9)] applied direct heating of the floating zone by induction heating with radio frequency so that in late 1954 the principles of the floating-zone technique for silicon were established. However, the dimensions of early float-zone crystals were small and their purity and quality of the latter only poor. A very important presupposition for a further development of the silicon float-zone method was an economical supply with starting polycrystalline silicon rods by deposition of silicon from gaseous mixtures of $SiHCl_3$ (trichlorosilane) or $SiCl_4$ (silicon tetrachloride) and hydrogen on thin, hot silicon cores at temperatures above 1000°C. The resulting polyrods could be used directly for the floating-zone process. By slim-rod pulling, new cores for further deposition of polycrystalline rods could be provided[10, 11)]. This so-called Siemens process, basically invented by Bischoff[12)], is the preferred method for semiconductor silicon production until today. A method suitable for silicon production was developed in the early fifties by Reuschel and coworkers[13)]. The chemical vapor deposition (CVD) process for the generation of hyperpure metals had been developed already by Van Arkel[14)]. Several workers utilized this method for silicon production[15, 16)]. However, all previous processes did not succeed because of insufficient purity and lack of economy.

At present, the following steps are usually involved in the production of float-zone silicon:

- Generation of slim rods (cores) by slim-rod pulling or by slicing of polycrystalline silicon.
- Mounting of the cores within the CVD reactor.
- Deposition of polycrystalline silicon from trichlorosilane and hydrogen above 1000 °C.
- Obtained polycrystalline rods are usually ground round for float-zone application. For Czochralski purposes, the polymaterial is grown thicker and brought into the crucible as lumps.
- Insertion of the polycrystal feed rod into the float-zone apparatus.
- Generation of a floating zone at the lower end of the feed rod and fusion with the seed crystal.

- Passing of the floating zone through the feed rod and rotating the growing single crystal in order to generate a straight and round single crystal.
- Doping of the crystals before, during or after single crystal growth.

These basic steps have not changed since they were first established in the fifties. However, the dimensions of the resulting crystals have grown by about one order of magnitude, the diameter from 10 to 100 mm and the length from 15 to 150 cm. Figure 1 shows a dislocation-free modern crystal. The crystal structure and the purity of recent float-zone crystals have also been improved greatly.

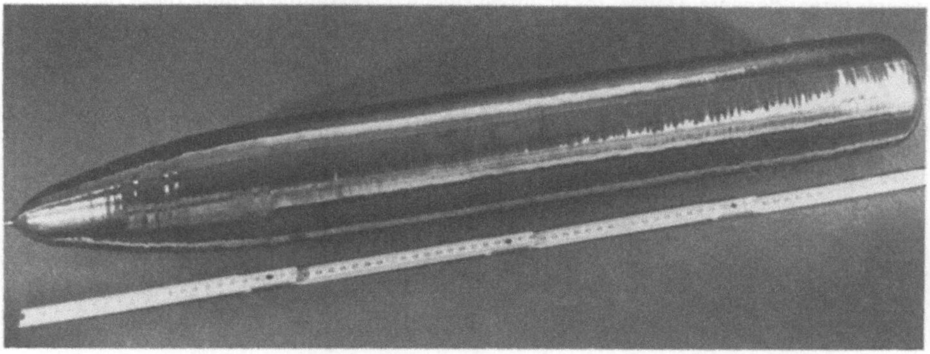

Fig. 1. Dislocation-free [100] silicon crystal of 104 mm diameter

Figure 2 shows the basic features of silicon float-zone growth. The floating zone is generated by means of a water-cooled induction coil fed with radio frequency (r.f.) power in the megacycle range. Because semiconductor silicon has only a small conductivity at room temperature, it must be preheated by radiation heating. When the crystal is hot enough it can pick up energy directly from the coil-field. The lower end is now molten and the seed can be introduced into the melt. By means of a relative motion between coil and crystal, either by moving the crystal downward or the induction coil upward, the floating zone passes through the feed rod starting from the seed. Both crystal ends,

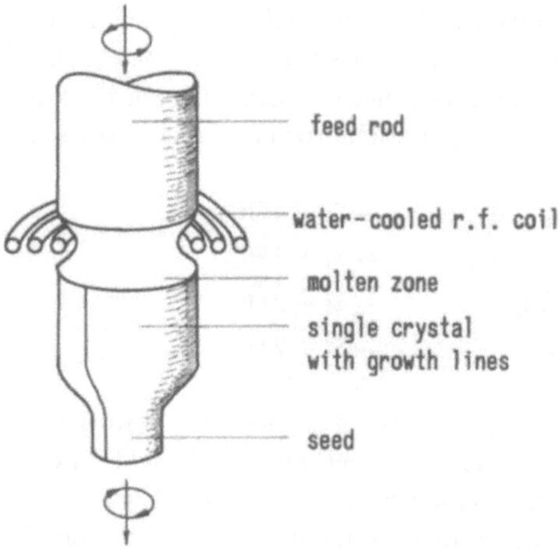

Fig. 2. Principle of float-zone crystal growth

particularly the seed end, are rotated, in general in opposite directions. This ensures growth of straight and round crystals. Growth can take place either in inert gas or in vacuum. The process can be repeated in order to increase the purity by the zone-refining effect. In vacuum, also the evaporation of volatile impurities is used for purification. The diameter of the resulting crystal is given by the growth rate and the amount of supply material which is fed into the floating zone. The floating-zone process can also take place downward with a hanging seed crystal, but this method has several disadvantages[1] and is used only for slim rod pulling. As already mentioned, the purification effect of floating-zone melting for most impurities is considerable. Exceptions are boron and phosphorus, two important dopants. Hence, for undoped silicon it is essential to have these two elements in only low concentrations in the starting material. Despite the zone-refining effect, a maximum purity is only achievable with a very pure starting polycrystalline silicon. This can be prepared only with hyperpure trichlorosilane and hydrogen.

2 Technical and Economic Aspects

The float-zone process is only one step, although an important one, in semiconductor silicon production. Figure 3 shows this step and where it fits into the full sequence of steps leading to semiconductor-grade silicon slices, starting from carbon and quartz sand[17]. Carbon and quartz sand are fused together in an electric arc furnace. Metallurgical-grade raw silicon is obtained with a silicon content of about 98%. Most of this silicon goes into metallurgical processes and to the chemical industry; only a small portion is needed for the production of semiconductor-grade silicon. Raw silicon is converted to trichlorosilane by reaction with hydrogen chloride in a fluid bed reactor at about 300 °C:

$$Si + 3\,HCl \leftrightarrows SiHCl_3 + H_2.$$

After distillation, trichlorosilane is reduced with hydrogen by the chemical vapor-deposition process. The obtained polycrystalline silicon now has a purity of 99.9999% as compared to 98% in raw silicon. As a by-product, silicon tetrachloride is formed. Usually, this substance is used for other chemical processes. However, it can also be reduced to silicon such as trichlorosilane, dichlorosilane and monosilane:

$$\begin{aligned} SiCl_4 + 2\,H_2 &\rightarrow Si + 4\,HCl \\ SiHCl_3 + H_2 &\rightarrow Si + 3\,HCl \\ SiH_2Cl_2 &\rightarrow Si + 2\,HCl \\ SiH_4 &\rightarrow Si + 2\,H_2. \end{aligned}$$

However, because of economical reasons, only trichlorosilane plays an important role in the production of high-purity silicon. Till now, the Czochralski method has mainly been applied to the production of single crystals where the crystals are pulled out of a crucible, in spite of their limited purity. It can be stated, however, that the float-zone method is

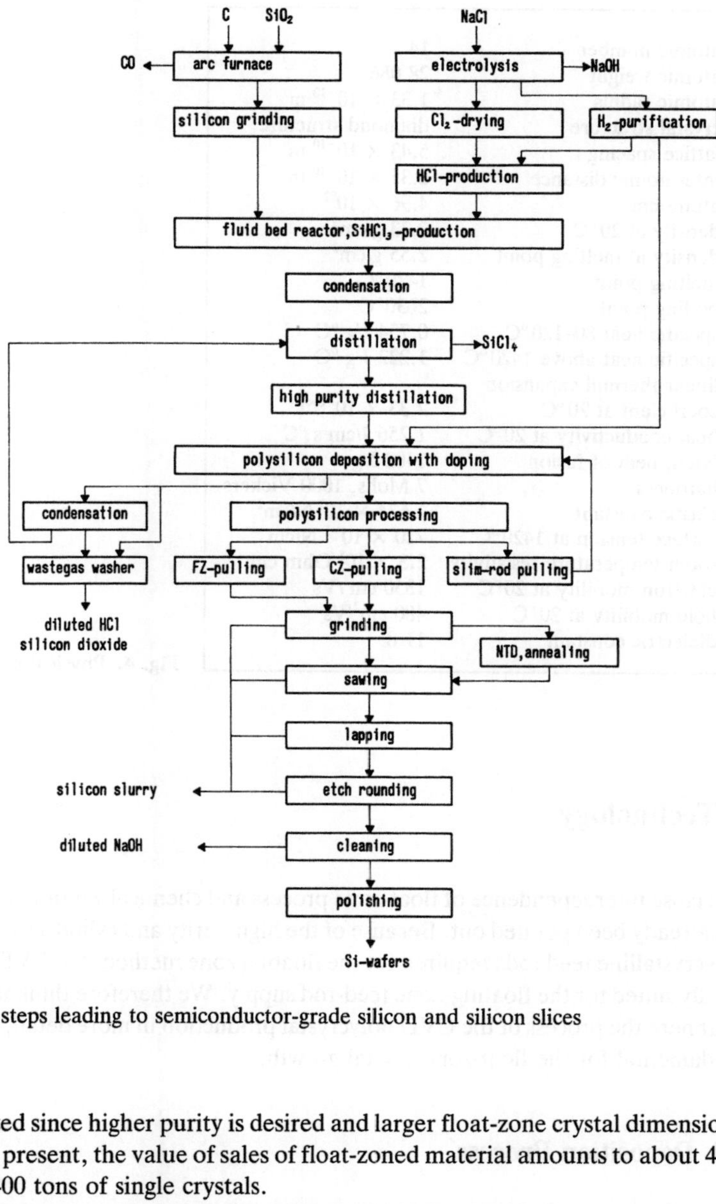

Fig. 3. Production steps leading to semiconductor-grade silicon and silicon slices

increasingly utilized since higher purity is desired and larger float-zone crystal dimensions can be grown. At present, the value of sales of float-zoned material amounts to about 400 million DM for 400 tons of single crystals.

The semiconductor-device industry needs monocrystalline silicon wafers with a thickness ranging from 0.2 to 1.5 mm. They are produced by slicing of the cylindrical silicon crystals with diamond saws. About half of the material is wasted here as slicing loss. Preferred wafer diameters today are 75 and 100 mm. Prior to further processing, the wafers are lapped, cleaned and polished, if necessary. High quality and low tolerances are required in these wafers because they are the base for high-quality device production. Because of its specific properties and the quality in which it can be produced, silicon has outstripped by far the other semiconductor materials. Figure 4 shows some important properties of silicon.

atomic number	14
atomic weight	28.086
atomic radius	1.33×10^{-10} m
crystal structure	diamond structure
lattice spacing	5.43×10^{-10} m
interatomic distance	2.35×10^{-10} m
atoms/cm^3	4.96×10^{22}
density at 20 °C	2.33 g/cm^3
density at melting point	2.55 g/cm^3
melting point	1420 °C
boiling point	2630 °C
specific heat 80–120 °C	0.754 J/g °C
specific heat above 1420 °C	1.222 J/g °C
linear thermal expansion coefficient at 20 °C	2.33×10^{-6}/°C
heat conductivity at 20 °C	1.256 J/cm s °C
latent heat of fusion	1402 J/g
hardness	7 Mohs, 1000 Vickers
elastic constant	1.674×10^{7} N/cm^2
surface tension at 1420 °C	720×10^{-5} N/cm
room temperature resistivity	2.3×10^{5} Ohm cm
electron mobility at 20 °C	1350 cm^2/Vs
hole mobility at 20 °C	480 cm^2/Vs
dielectric constant	11.8

Fig. 4. Physical properties of silicon

3 Technology

The close interdependence of float-zone process and chemical vapor deposition of silicon has already been pointed out. Because of the high purity and cylindrical size of the polycrystalline feed rods required for the floating zone method, the CVD method is ideally suited for the floating-zone feed-rod supply. We therefore think it necessary to treat here the process of the CVD polycrystal production in more detail, because it is so fundamental for the float-zone crystal growth.

3.1 Deposition Process

Figure 5 shows the principle of a CVD-reactor designed for the deposition of polycrystalline silicon rods. The deposition of silicon takes place on silicon surfaces which have a temperature of 1000 °C and more. The active surface for deposition should be as large as possible in order to allow for high deposition rates which are economically essential. The choice of materials for the construction of the reactor is rather critical because of the high temperature, at which no impurities from the walls should be introduced into the system in a highly corrosive environment which contains hydrogen chloride. Quartz glass is best suited for the bell material which encloses the reaction system. The quartz bell is mounted tightly on a water-cooled silver base plate. In order to reduce heat loss, the base plate

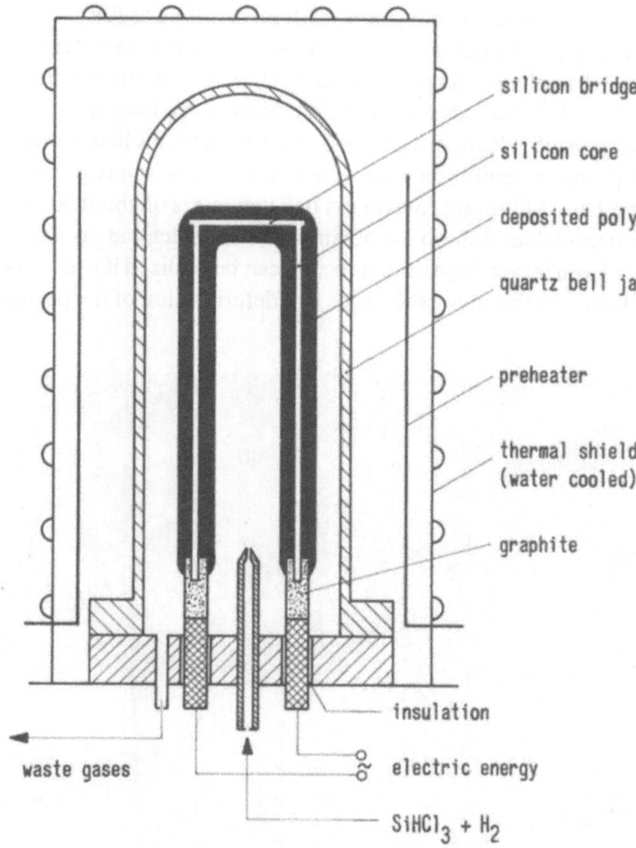

Fig. 5. Deposition reactor for preparing polycrystalline silicon rods (schematic)

is covered with quartz plates. The base plate has gas-tight insulated metal-graphite fittings for the silicon cores on which deposition takes place. The two core tops are connected by silicon bridges. The base plate has a gas inlet for the reactive gas mixture of trichlorosilane and hydrogen. The decomposed gas mixture leaves the vessel through the gas outlet. The reaction chamber is surrounded by a water-cooled metal bell which picks up the heat coming from the reactor and ensures a sufficiently low temperature of the quartz bell in order to avoid decomposition here. Prior to starting the reactor, the silicon cores must be heated up so that their electrical conductivity becomes sufficient for direct current heating. This can be achieved by radiation of a resistance heater from the outside. Another way is to heat the cores up to the necessary temperature of 300 °C by high voltage of several kV. However, problems with the insulation of the electrodes may arise here. As soon as the core conductivity is high enough, further heating takes place by current flow. Because of its negative temperature coefficient, the resistivity of silicon decreases further until the reaction temperature of 1000 °C and more is reached. For a reliable electrical power control, it is essential to keep the current stable as otherwise no constant deposition temperature is maintained. This is achieved by means of a thyristor-controlled power supply. The pertaining power transformer must be suitable for the handling of largely different working conditions. As long as the growing polycrystals are thin, relatively high voltages are necessary. With growing diameters and for a constant temperature, the voltage drops considerably ($U \sim d^{-1/2}$) whereas the current increases drastically ($I \sim d^{3/2}$). Of course, deposition reactors can have more than only two silicon cores. Reactors with 4, 6, 8, or 12 rods are useful, two rods being always connected by a silicon bridge. A size limitation of the reaction chamber is only given by the limited size of the quartz bell. At present, maximum bell diameters of about 500 mm at a height of 2000 mm are available. Bells made of vitreous fused silica can even have somewhat larger dimensions. Even larger deposition reactors can be realized if water-cooled metal bells are used. However, this inevitably leads to a deterioration of the purity of the growing

Fig. 6. Six-rod deposition reactor after finished silicon deposition reaction

polycrystals. Hence, for float-zone application, bells made from quartz glass are preferred. Figure 6 shows a CVD reactor with 6 as-grown polycrystalline silicon rods. Note the straight and almost cylindrical shape of the rods which makes them ideally suited for float-zone growth.

As already mentioned, the following reaction takes place in the reactor

$$SiHCl_3 + H_2 \leftrightarrows Si + 3\,HCl.$$

The equilibrium of the reaction is temperature dependent. Deposition of silicon occurs at high temperatures whereas the reverse reaction proceeds at low temperatures. As we have already seen, trichlorosilane is prepared from silicon and hydrogen chloride at lower temperatures in the fluid bed reactor.

Several side reactions, which are also temperature dependent, occur in a CVD reactor, e.g.

$$SiHCl_3 + HCl \leftrightarrows SiCl_4 + H_2$$
$$SiHCl_3 \leftrightarrows SiCl_2 + HCl.$$

The latter reaction gives silicon dichloride which is only stable at high temperatures. It polymerizes at lower temperatures to form polysilanes ($[SiCl_2]_x$) which are very reactive, even explosive, viscous liquids which obtain in small quantities during the deposition process. Because of the formation of silicon dichloride, the yield of silicon cannot be improved by means of deposition temperatures above 1200 °C. The possible equilibrium reactions of the system Si − H − Cl have been extensively studied[18,19]. One could keep in mind, however, that equilibrium conditions are hardly obtained in a reactor for high deposition rates. Here, kinetic factors play a dominant role. For the actual process, some interdependences are important which are not contradictory to the points mentioned before. At constant deposition temperature, the increase of gas flow will lead to a higher silicon deposition rate, yet a lower silicon yield with respect to trichlorosilane consumption. At constant gas flow, an increasing molar ratio of the gas mixture $SiHCl_3/H_2$ also leads to a higher deposition rate and lower silicon yield. Increasing temperature or larger deposition surface, i.e. larger rod diameter, raises deposition rate and silicon yield simultaneously. However, for the latter case, the energy consumption and the thermal stress of the reactor are enlarged. If the surface temperature is raised, the surface structure of the growing rods is deteriorated because the surface crystallites grow larger and more uneven. In this case, the rods must be ground round and smooth for float-zone application. A loss of material is inevitable. Optimum growth conditions must be chosen to meet the desired economic and technical requirements. Factors which are important here are gas flow, molar ratio, temperature, deposition rate, yield, size, and quality of silicon surfaces and, if the waste gases are recycled, also their composition. It is obvious that during a run, the deposition conditions must necessarily change because of the variation in diameter. Hence, it is useful to adapt the remaining parameters to the changing growth conditions during a run. Since the silicon deposition rate in a CVD reactor hardly reaches 1 mm/h, the process is rather slow and energy-consuming. For float-zone applications, rods with diameters up to about 100 mm and lengths up to 1500 mm are required. On the other hand, diameters up to 250 mm are useful for Czochralski growth. Then, a run will require many days.

The quality requirements for Czochralski polycrystalline feed material are limited both by purity and structure. The rods need not be crack-free because generally lumps are filled into a crucible. Much higher quality levels are necessary for float-zone feed rods. Their cross section should be nearly circular. Moreover, the crystallites should be small so that a smooth surface is approached (Fig. 7). Often, it is easier and cheaper to grind than to grow a smooth and cylindrical surface, especially for larger rod diameters. Cracked crystals cannot be processed because it is difficult to melt them during the floating-zone process and the crystal may break. Gas inclusions can lead to explosions during melting, especially under vacuum very harmful electrical discharges in the coil-melt region may occur. The first of all requirements for undoped feed material is high purity. The resistivity should exceed 1000 Ω cm. However, this can be measured only in the single crystalline material and not in the polymaterial because of its grain boundaries. The dopants boron and phosphorus should be present only in concentrations below 5×10^{12} at/cm^3. A low boron concentration is particularly important because boron cannot be removed during the floating-zone process. Carbon and oxygen are usually found in somewhat higher amounts, yet they should not exceed 10^{16} at/cm^3. Polyrods which meet these stringent specifications can be made only from pure trichlorosilane and hydrogen with particularly low boron and phosphorus contents. For the floating-zone process not only undoped silicon is desired. Feed material doped to a certain dopant level of e.g. boron or phosphorus is also useful. This doping can be readily achieved during deposition if the gaseous mixture of trichlorosilane and hydrogen is doped with a controlled amount of dopant, e.g. phosphorus trichloride or boron tribromide[20)]. The method is not very accurate and only suited for higher dopant concentrations as the amounts of dopant for deposition are small and an exact dosage is difficult to apply. Slim-rod doping gives a higher precision of doping. Here, exactly measured doped cores are used for the deposition of undoped silicon. If the polyrod grows to the calculated diameter, one obtains fairly exact dilution parameters and even low doping levels in the feed material can be

Fig. 7. Polycrystalline silicon rod

reproduced rather well. In both CVD doping processes, the dopant is not distributed uniformly over the cross-section of the as-grown polyrod. The necessary homogenization takes place during the following floating-zone melting. Very low dopant levels can be achieved by means of the generation method[21] (Fig. 8). The starting rod is produced by deposition doping as described above. Then, by successive and repeated steps of slim-rod pulling, deposition and floating zone, the desired dopant level in the final single crystal results. Of course, it is essential for the process that the polymaterial deposited during the CVD steps is sufficiently pure.

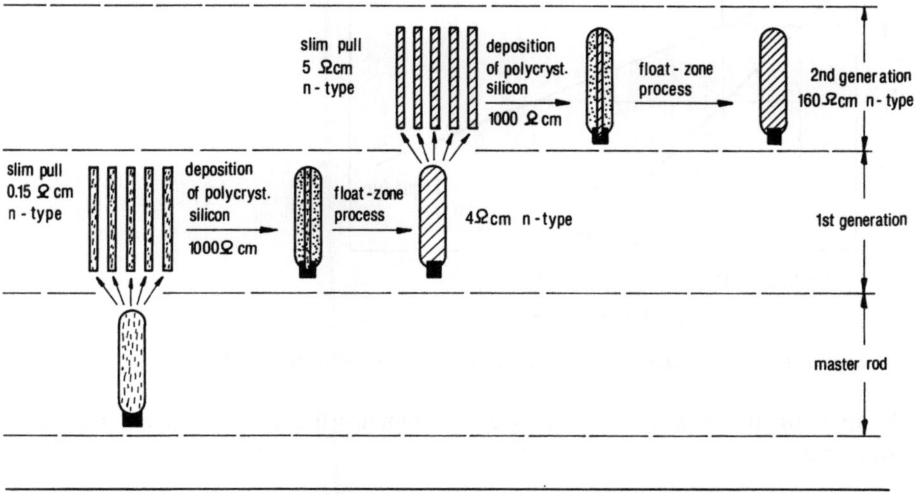

Fig. 8. Generation doping method

3.2 Floating-Zone Process

During float-zone growth, two basic effects become operative. Firstly, purification by zone refining and secondly conversion into a single crystal. The optimum growth conditions usually deviate more or less from optimum purification conditions. Hence, it can be useful to separate both, e.g. by starting with one or several purification passes under optimum purification conditions and thereafter growing the single crystal under optimum crystallization conditions. Because of the purity of the polymaterial achieved today, the emphasis of the float-zone process now certainly lies on single-crystal conversion under optimum conditions. Purification by zone refining does no longer play the dominant role which it had in the early times of float-zone silicon crystal growth where only poor feed material was available.

3.2.1 Zone-Refining Conditions

In his zone-refining work, Pfann made use of the fact that homogeneously distributed impurities in a slowly freezing melt, where the solid-liquid interface moves, usually segregate. In a freezing silicon melt, almost all impurities prefer to stay in the melt, i.e. this melt will pick up more and more of the impurities during normal freezing. Thus, the impurity concentration both in the melt and in the growing solid will increase as solidification proceeds. This follows from the solidus/liquidus curves of the two soluble substances, e.g. silicon and an impurity (Fig. 9). For small impurity concentrations, these curves

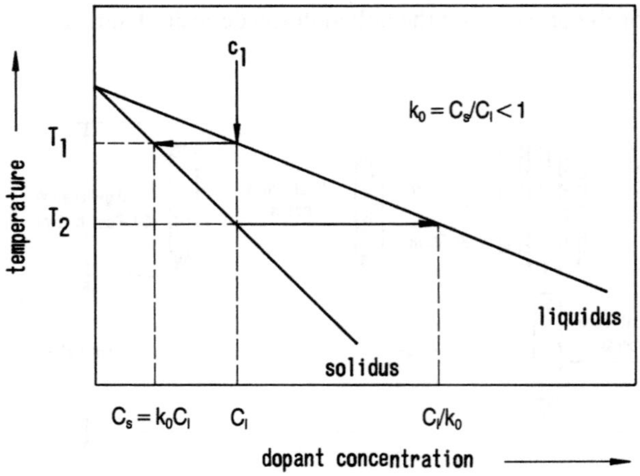

Fig. 9. Solidus-liquidus diagram for a dopant element in silicon with $k_0 < 1$

can be approximated by straight lines. An equilibrium distribution coefficient k_0 can be derived where

$$k_0 = \frac{C_s}{C_l}$$

C_s = impurity concentration in the solid
C_l = impurity concentration in the liquid

For small concentrations, k_0 is independent of the impurity concentration. If a floating zone of length l slowly moves through a rod of length z and impurity concentration C_0, the impurity concentration at the zone start will be diminished by the factor $k_0 \times C_0$, i.e. the more, the lower k_0. The impurity concentration increases along the rod; where the floating zone freezes, a particularly high impurity concentration occurs. This purification process can be repeated until a final concentration C_z is reached

$$\frac{C_z}{C_0} = A \exp\left(B \cdot \frac{z}{l}\right)$$

where A and B are constants which contain k_0. This equation shows that the impurity depletion also depends on the ratio z/l. A large crystal length z and small zone length l permit a smaller final concentration near the zone start. The final distribution after many zone passes is independent of the crystal diameter. The above discussion is only valid for equilibrium conditions, i.e. if the growth rate is so low that full mixing of the zone by diffusion is possible. Under actual growth conditions, the growth rates are so high that an equilibrium no longer exists. The growing interface migrates so rapidly that the impurity atoms in the melt will dam up forming a diffusion layer with enlarged impurity concentration at the growing interface. k_0 still describes the concentration change in the liquid-solid transition but no longer the segregation of the impurity content in the whole molten zone. This is expressed now by the effective distribution coefficient

$$k_{eff} = \frac{C_s}{C_l}$$

and

$$k_{eff} > k_0$$

for all impurities with $k_0 < 1$.

Figure 10 shows k_{eff} as a function of the growth rate and Fig. 11 the equilibrium distribution coefficients k_0 for impurities in silicon. It is obvious that boron is the element

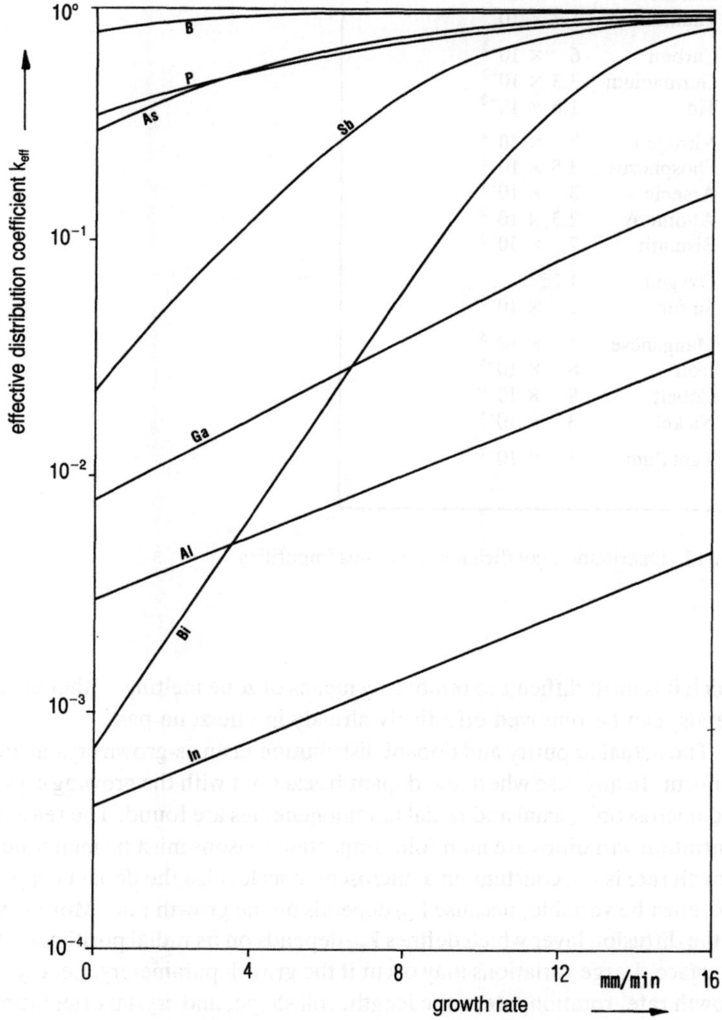

Fig. 10. Dependence of the effective distribution coefficient on the growth rate for several dopant elements (calculated)

Element	Distribution Coefficient k_0
Lithium	1×10^{-2}
Copper	4×10^{-4}
Silver	1×10^{-6}
Gold	2.5×10^{-5}
Zinc	1×10^{-5}
Cadmium	1×10^{-6}
Boron	8×10^{-3}
Aluminum	2×10^{-3}
Gallium	8×10^{-3}
Indium	4×10^{-4}
Thallium	1.7×10^{-4}
Carbon	6×10^{-2}
Germanium	3.3×10^{-1}
Tin	1.6×10^{-2}
Nitrogen	7×10^{-4}
Phosphorus	3.5×10^{-1}
Arsenic	3×10^{-1}
Antimony	2.3×10^{-2}
Bismuth	7×10^{-4}
Oxygen	1.25
Sulfur	1×10^{-5}
Manganese	1×10^{-5}
Iron	8×10^{-6}
Cobalt	8×10^{-6}
Nickel	3×10^{-5}
Tantalum	1×10^{-7}

Fig. 11. Distribution coefficients of various impurities

which it is most difficult to remove by means of zone melting. Other elements, e.g. most metals, can be removed effectively already in one zone pass.

The actual impurity and dopant distribution in an as-grown crystal are generally non-uniform. In any case where the dopant freezes out with the growing crystal, macroscopic and microscopic, axial and radial inhomogeneities are found. The reason for these concentration variations are manifold. Important reasons must be mentioned here. Since the growth rate is not constant on a microscopic scale, also the dopant (impurity) concentration must be variable, because k_{eff} depends on the growth rate. Moreover, the thickness of the diffusion layer which defines k_{eff} depends on its radial position on the growing interface. Large variations may occur if the growth parameters, i.e. crystal diameter, growth rate, rotation rate, zone length, coil shape, and crystal orientation are not selected properly[1)]. Certainly, problems arise here because the required crystal quality may grow only under conditions where the dopant segregation is not uniform.

3.2.2. Crystallization Conditions

The float-zone crystallization process involves a much higher number of parameters than the simple original zone-melting technique in a boat. One cause is that the molten zone is generated by induction heating. The most important stabilizing factors of the floating zone are the relatively high surface tension (720 mN/m) and low density (2.55 gcm^{-3}) of molten silicon. Both factors greatly help to hold the molten zone between feed rod and single crystal. Of course, there exists a maximum zone length above which the floating zone would no longer be stable. Brice[22)] has described this relationship in more detail. Experimental results show that the external length of the floating zone should not exceed 30 mm[1)].

Additional support for the molten zone comes from the electromagnetic coil field. It gives rise to electrodynamic supporting forces which have the same order of magnitude as the surface tension forces[1)]. Hence, the molten zone can be made considerably longer and more voluminous than with surface-tension-driven support only. This is particularly important for the growth of larger crystal diameters under non-needle-eye growth conditions (see Sect. 3.2.3). Here a full melt-through of the feed rod can only be realized with a larger external zone length. The solid-liquid interface tension also increases stability, although the resulting forces are not as strong as those mentioned above. In this context, the shapes of the growing interface at the bottom of the zone and the melting interface at its top must be considered. Positive effects are likely to come from the actual technical shapes of interfaces with the growing interface being concave, i.e. depressed, the melting interface convex, i.e. extending into the floating zone[1)].

The most important destabilizing force on a floating zone is gravity. Its negative influence is obvious in large diameter crystal growth. Hence, without gravity one should be able to grow any diameter which, in fact, may be a future task for special purposes. During the floating zone process, generally both crystal ends are rotated. The rotation of both interfaces generates centrifugal forces in the adjacent melt which bring about fluid flow and influence the stability of the melt. If a certain rotation rate is exceeded, the floating zone will spill over its lower interface. The critical rotation rate decreases with increasing crystal diameter. Vibrations can also lead to instability of the floating zone if their frequency is low and their amplitude high. Especially critical are resonance effects either of the melt directly or with the growing crystal plus melt. Vibration sources with frequencies between 1 and 50 Hz usually stem from the mechanical drives of a float-zone apparatus. The system consisting of thin seed, bottle neck and growing crystal is particularly sensitive to low-frequency vibrations. Additional supply systems are necessary for larger crystal weights in order to avoid loss of the melt as a consequence of strong vibrations. Vibrations can also come from the melt-coil field, if the r.f. power supply has too much ripple. Well filtered direct current is necessary for the r.f. generator power supply.

The convection flow in the melt must also be taken into account. Its influence on melt stability is only slight, but crystal growth, i.e. dopant and defect build-up, and shapes of the interfaces strongly depend on it. At least four different kinds of convection flow must be distinguished. Thermal free convection flow is brought about by temperature differences in the melt which cause density differences. The surface of the melt is hottest in the coil plane. Light hot melt flows up from there to the feed rod, cools down by melting up feed material and sinks down to the growing interface in the inner melt (Fig. 12). This

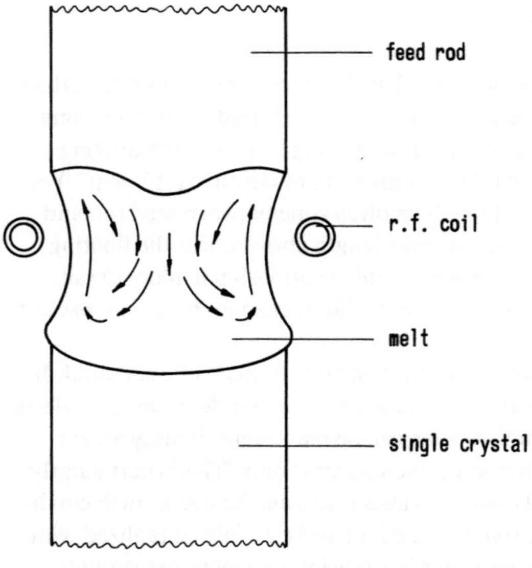

Fig. 12. Thermal-free convection in a floating zone

effect can be strongly enhanced if the growing interface is rotated. The centrifugal flow generated has very roughly the same direction but can be far stronger than thermal flow[23]. Without doubt, the electrodynamic forces of the coil also lead to convection flow. In the coil plane it is directed radially inward. Two toruses may result which close by a flow radially outward along the interfaces. Also asymmetric toruses with secondary vortices may occur. Figure 13 shows possible flow patterns for a symmetric arrangement of coil and melt. Experimental evidence of this pattern has not yet been established. The Marangoni effect should cause another, i.e. surface-tension driven flow[24]. This effect

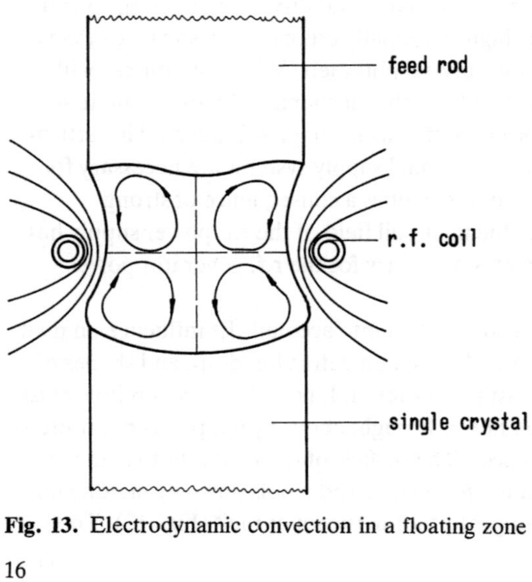

Fig. 13. Electrodynamic convection in a floating zone

results from the temperature dependence of surface tension. The flow always goes from lower to higher surface tension, i.e. with decreasing surface temperature. Figure 14 shows the flow pattern to be expected which is opposite to the electrodynamic pattern so that some compensation may obtain. Experimental evidence of Marangoni flow on silicon melts is still lacking too.

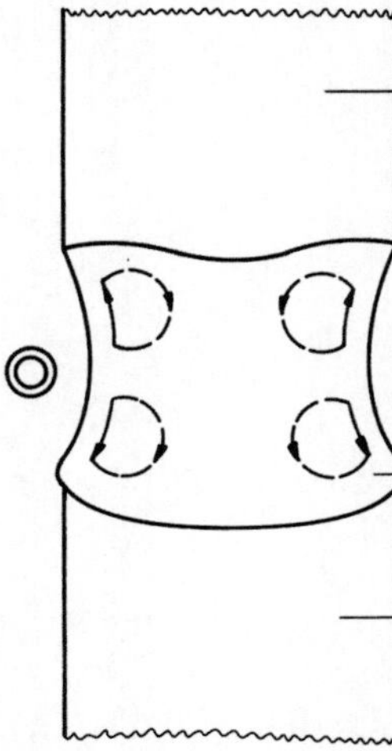

Fig. 14. Marangoni flow in a floating zone

3.2.3 Floating-Zone Equipment

Modern float-zone machines must meet many stringent specifications of rather different kind. They must be highly stable mechanically in order to ensure a smooth, vibration-free growth. Only smoothly running drives are permitted. All feed-through sleeves must be high-vacuum tight to permit operations under vacuum. The current feeding connection to the melt coil is a particularly critical part which has to carry reliably high r.f. current densities. It is obvious that modern control equipment is also involved.

Two concepts for floating-zone machines are useful. In the first one, the crystal rod is fixed whereas the coil is movable while in the second one, the silicon rod is passed through a fixed coil. For both concepts, the drives can be mounted within or outside of the working chamber. Although movable coil zoners need less clearance, fixed coil zoners offer distinct advantages for large-diameter crystal growth where high melt-power levels must be fed to the induction coil. Modern fixed coil zoners with a growth length of 1500 mm can require a room clearance up to 9 m (Fig. 15). Economic considerations require silicon crystal rods with large diameter and length to increase the volume output. Large rod lengths reduce the portion of the waste crystal ends with respect to the ingot weight.

The production of large dislocation-free crystals requires supporting systems, as only crystal weights up to 4 kg can be carried directly by a thin seed with bottleneck. Figure 16 shows such a supporting system which allows the growth of larger weights. Here, the crystal cone is supported axially and radially[25, 26].

Fig. 15. Fixed-coil float-zone apparatus for large large crystal diameters

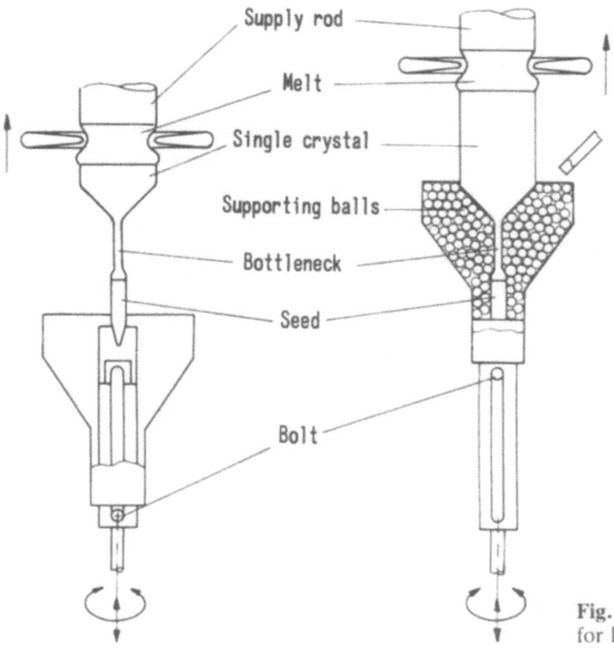

Fig. 16. Crystal-supporting system for heavy crystals

Heating of the floating zone takes place by means of r.f. eddy currents generated by the magnetic coil field. For crystal diameters above 50 mm, a working frequency between 2 and 3 MHz is useful[1, 27]. Higher frequencies give rise of electric arcing which is not tolerable. Moreover, they reduce the desired supporting action of the coil field. On the other hand, high working frequencies would be advantageous for the melt-up of the feed crystal surface grains which do not melt properly with too low generator frequencies. Higher frequencies also make it easier to fuse seed crystal and feed rod and to grow a bottleneck. Argon which is the favored inert gas for float-zone growth has a relatively low electric breakdown voltage. Therefore, the melt coil is designed in a way which allows operation under low voltage. As a consequence, the current flowing in the coil is rather high in order to reach the necessary power levels. Here, one-turn, low-inductance water-cooled coils with an internal diameter between 30 and 38 mm are useful. They can be made separable[28] where the separation point is grounded[29]. This leads to a reduction of the voltage between grounded melt and coil feed ends by a factor of two.

With the modern needle-eye technique for large-diameter crystal growth, the thick feed rod melts at a narrow melt coil. After passing this needle-eye the single crystal grows thick again[30, 31]. When growth is finished, the coil can be opened and the crystal can be removed from its holders.

A first warming up of the feed rod prior to melting is necessary because cold silicon has a too low conductivity to pick up energy directly from the coil. This is achieved by means of a glowing graphite ring which can be turned away after heating up.

The design of an induction heating system used by Siemens since many years is explained in Fig. 17. The system consists of an r.f. generator with variable frequency

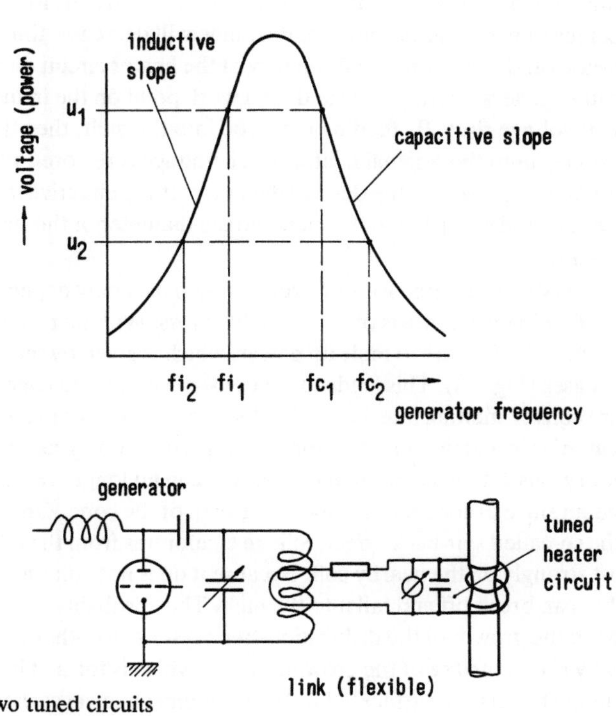

Fig. 17. Heater system with two tuned circuits

which is coupled with a tuned heating circuit set close to its resonance frequency. The link cable between generator and heating circuit can be a flexible coaxial cable as it has to carry only a fraction of the heating circuit current. By proper choice of the generator frequency one can either work on the inductive or capacitive slope of the heating circuit resonance curve. The desired voltage u_1 which defines the heating power can be reached on the inductive slope with generator frequency f_{i_1} and on the capacitive slope with generator frequency f_{c_1}.

With increasing generator frequency, the work point on the inductive slope goes up, on the capacitive slope down, respectively. At constant generator frequency, a work point on the inductive slope indicates thermal self-stabilization[32]. The reason is that the resistivity of silicon decreases with increasing temperature, leading to a decreasing melt-coil inductance and consequently rising heating circuit resonance frequency. The resonance curve (Fig. 17) moves to the right, the coil voltage and thus heating power decreases and dangerously fast heating-up of a silicon rod can be avoided. On the other hand, a work point on the capacitive slope indicates mechanical self-stabilization of melt and crystal[1]. If crystal and melt bend toward the inner coil rim because of vibrations or gravity, the coil inductance decreases. Consequently, the resonance curve is shifted to the right and the coil voltage and power increase resulting in an increase of electrodynamic forces. These forces push melt and crystal back to the coil center, i.e. rotation axis. This effect is rather important for the growth of large-diameter crystals. With the two-circuit power supply system described, a simple diameter control by means of stretching and squeezing of the floating zone can be realized[33]. By simply feeding more or less feed rod to the melt coil its inductance and thus the voltage of the heating circuit can be kept constant. If this is performed with a closed-loop control system, the resulting automatic diameter control gives about 1% diameter uniformity. If, for instance, the feed rod becomes thinner, the effective melt volume will decrease; thus, the coil inductivity increases and the resonance frequency of the heater circuit diminishes. This means that at constant generator frequency and for a work point on the inductive slope, the heater circuit voltage rises. By feeding more rod into the melt, the effective melt volume increases, until the original heater circuit voltage is restored whereby the single crystal continues to grow with the desired diameter. If the effective melt volume becomes too large less feed rod is fed to the melt and the diameter of the single crystal again remains constant.

The power required for float-zone growth naturally depends on the crystal diameter[1]. It predominantly consists of radiation heat loss. With increasing crystal diameter, the ratio of radiation loss to melt-up power, which is given by the latent heat of fusion, decreases (Fig. 18). This leads to a more strongly curved concave growing interface and a more convex melting interface. The float-zone process can either be performed under vacuum or in an inert gas atmosphere. It has been observed, however, that dislocation-free crystals with a diameter above 30 mm can no longer reproducably be grown under vacuum since dislocations arise after growth of the cone. One cause for this phenomenon is the so-called spit-back effect. Silicon evaporates from the melt surface and is deposited most strongly on the nearby coil. Because it does not adhere tightly to the latter, small flakes can break off and fall into the melt. They probably reach the growing interface and disturb the growth of the dislocation-free crystals. Another reason may be the microscopically rough surface of the growing crystal which is formed by evaporation of the melt from the hot crystal surface[1]. Under thermomechanical stress this rough surface is likely

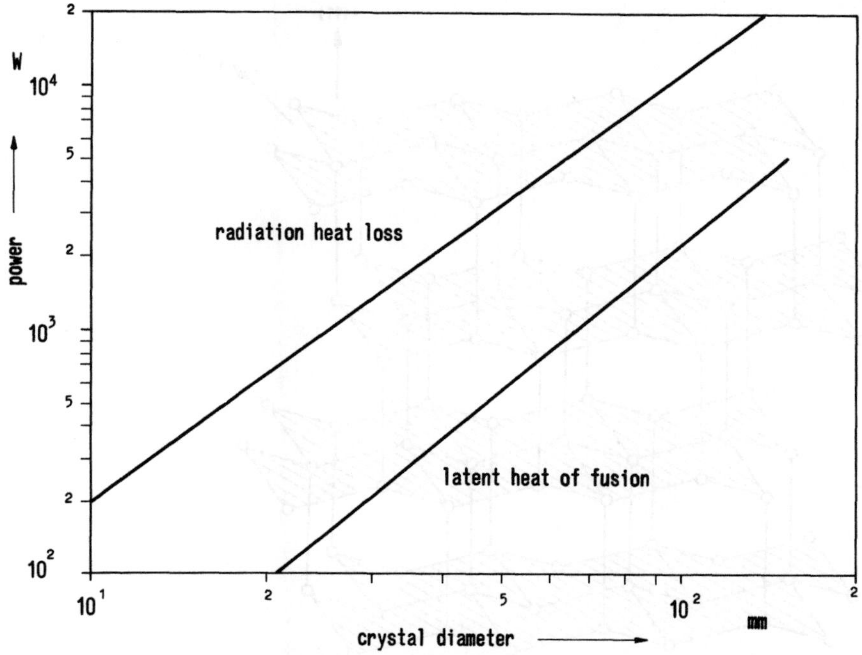

Fig. 18. Calculated radiation heat loss and latent heat of fusion for different diameters

to generate dislocations the number of which increases in the bulk crystal by multiplication. Argon is the preferred inert gas for the floating zone technique because of its low price and sufficient breakdown voltage. Under sufficient pressure (about 1 bar), the above mentioned difficulties are avoided because silicon can no longer evaporate from the melt and the hot solid. Helium is more expensive and has a too low breakdown voltage. Hydrogen would be very suitable but gives rise to undesired lattice defects. The argon used must, of course, have an excellent purity.

3.2.4 Silicon Crystallography

For the understanding of the growth and properties of a silicon single crystal it is useful to know the basic crystallography. Silicon crystallizes in the cubic diamond lattice (Fig. 19) with space group Fd 3 m and shows no lattice transformation until melting. Each silicon atom has four nearest neighbors due to its eletron configuration. The silicon lattice is not very densely packed. As a consequence, solid silicon has a lower density than liquid silicon. The most densely packed crystal planes are the {111} planes. They are energetically favored and hence a freely growing silicon nucleus forms an octahedron (Fig. 20). A reference octahedron is quite useful for obtaining a rapid crystallographic survey, this allowing to easily find the important directions [111], [100], [110] and [112] with the corresponding crystal planes perpendicular to these directions. The [111] orientation has a trifold symmetry which leads to three lengthwise growth lines set at 120° along the periphery of a [111] crystal in the directions [$\bar{2}$11], [11$\bar{2}$] and [1$\bar{2}$1].

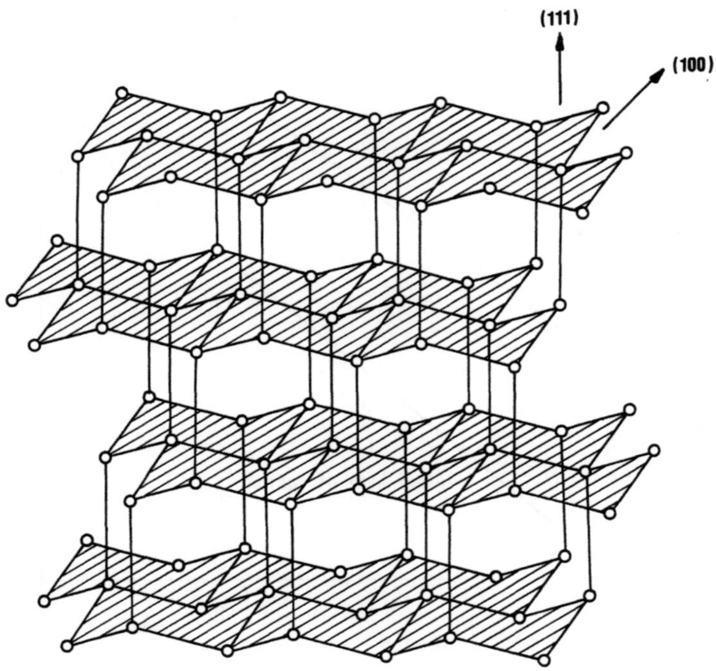

Fig. 19. Crystal lattice of silicon showing the {111} planes and the tetrahedral structure

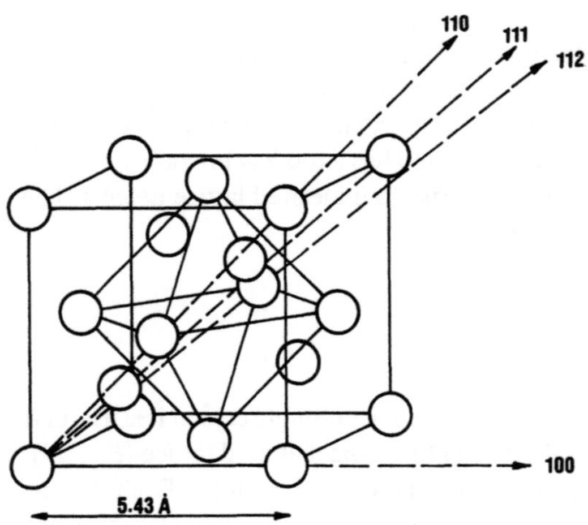

Fig. 20. Cubic cell of silicon with the octahedron and some important crystal directions

Etching on {111} planes results in trifold etching figures. A [100] crystal shows fourfold symmetry with four growth lines set at 90° in the directions $[0\bar{1}\bar{1}]$, $[0\bar{1}1]$, [011] and $[01\bar{1}]$. Each figures on {100} also display fourfold symmetry. For the [110] orientation, one obtains a twofold symmetry with twofold etch figures on {110}, and [112] crystals have a singlefold symmetry only. In order to grow a crystal in a desired orientation, it is necessary to adjust the seed crystal accordingly. The desired crystallographic direction and the rotation axis of the crystal must coincide because the crystal grows along the rotation axis. For extremely precise measurements of the orientation, an X-ray goniometer which measures the Bragg angles can be used. In general, however, an optical method is adequate[34, 35]. It has a precision of about $\pm 0.1°$ and allows for a simple and fast adjustment of a seed crystal. After caustic etching of, for instance, a [111] seed crystal, its end faces will show trifold etch pits. If one end face is illuminated by a narrow beam along the rotation axis, it will reflect a trifold star on an observation screen which resembles a Mercedes star. If the center of this star does not migrate on rotation of the seed, it is exactly [111]-oriented along the rotation axis. For a certain misorientation, the center of the reflexion star rotates around the rotation axis, the distance from the axis being a measure of the misorientation. A [100] seed orientation should reflect a fourfold pattern. However, the {111} reflexes are much more marked; hence, {111} reflexes are also used to adjust other orientations such as [100]. Here, the rounded and etched tip of a mounted [100] seed will reflect four equidistant {111} reflexes at an angle of 54.8° toward the rotation axis, because the angle between the directions [100] and [111] is 54.8°. Starting with a [111] seed, one can grow crystals of other orientations by properly tilting the [111] seed. This is described in more detail in[1].

3.2.5 Single-Crystal Growth

To grow a single crystal, the oriented seed crystal is inserted into the molten bottom tip of the feed rod. Since the floating zone is moved through the feed rod, a single crystal crystallizes under the floating zone. To secure an appropriate crystal quality, this single crystal should be dislocation-free. Since a seed always contains dislocations, a special technique is necessary to avoid the transition of dislocations into the growing crystal. According to the basic work of Dash[36], which has been adapted to technical requirements by Ziegler[37], dislocation seeding can be avoided by necking-in the generally used thin seed[38] over, a length of 20–30 mm at the fusion point of the seed. The necked-in portion has a diameter below 3 mm. If the growth rate in the bottleneck is chosen properly, the dislocations end here and one can build up even large diameters, dislocation-free on a dislocated seed. Several hypotheses on the annihilation of dislocations in the bottleneck have been proposed. This annihilation is presumed to involve a reaction of dislocations with point defects which shifts the dislocations to the surface where they disappear. Moreover, the necessary high growth rate in the bottleneck for [111] growth may exceed the maximum possible propagation rate of dislocations and this causes them to stop in the bottleneck. During stationary float-zone crystal growth, many factors which influence crystallization must be controlled (Fig. 21). It is obvious that for an economic floating zone growth process, the feed rod should have about the same length as the single crystal in order to utilize the full growth length of the zoning machine. However, one can also squeeze a thinner feed rod to a thick single crystal[30] or, vice versa, stretch a thick feed rod into a thinner single crystal[39]. The rotations of feed rod and single crystal are generally opposite. For large diameters, the crystal rotation rate must be low (below 10 rpm) as otherwise spiral growth or melt loss will occur because of too high centrifugal forces. Increasing crystal diameters cause rising macroscopic doping inhomogeneity over the cross section. This inhomogeneity can be reduced by growing a crystal from an

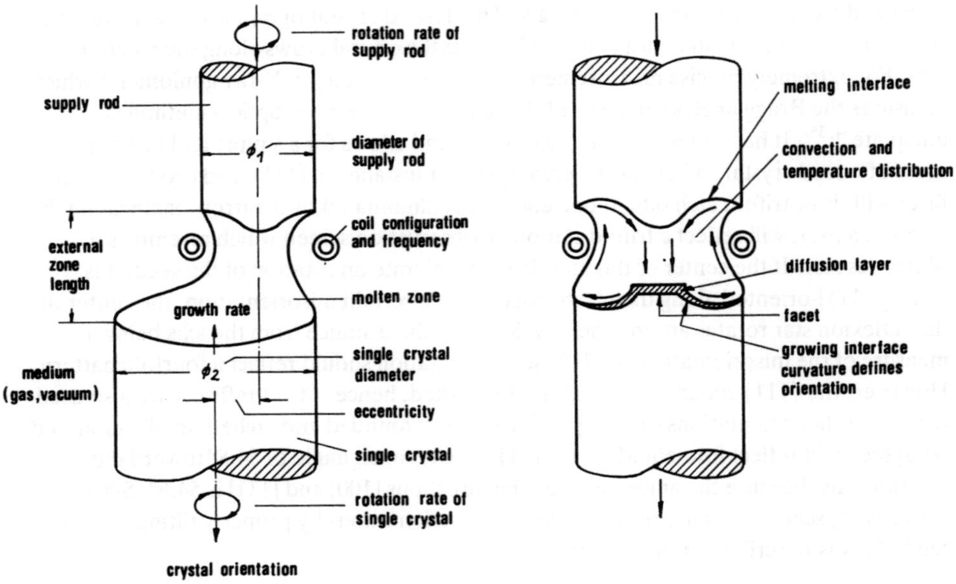

Fig. 21. Parameters which influence the float-zone process

eccentric position of the crystal with respect to coil and feed rod[40], however, only at crystal rotation rates of at least 10 rpm where strong centrifugal flow in the melt is effective. This means that the eccentric mode of operation is useful mostly for small and medium crystal diameters which can be rotated fast enough. Quite important is also the coil shape. For crystal diameters above 25 mm, the inner coil diameter must be smaller than the crystal diameter. The outside coil diameter should be larger so that the electrodynamic supporting force of the coil sufficiently centers the melt. The influence of r.f. induction heating has been mentioned already, also the effect of the zone length which cannot exceed 30 mm for the sake of stability. Since the heat of fusion per unit area is proportional to the growth rate, the growth rate has a considerable effect on the shape of the interfaces which increases with diameters. A further essential parameter of the float-zone growth is the environment, i.e. either vacuum or gas. All the parameters mentioned interdepend and must be adapted to the process[23]. Facet growth is observed only for crystals grown in the [111] direction. A facet is a smooth {111} plane element which generally grows on convex portions of the growing interface. The curved rough regions of the interface coincide well with the crystallization isotherm whereas faceted regions deviate from the latter and are usually supercooled. The reason is that {111} is the most stable crystal plane on which growth nuclei cannot form as easily as on the curved rough interface regions. The supercooling provides the necessary driving force for a nucleation on facets. Ciszek[41] has estimated that for the large central growth facets a supercooling of 5 °C is possible; hence, the deviation from the crystal isotherm is considerable. In faceted smooth interface regions, rapid lateral growth of successive lamellae occurs. This leads to an increased dopant incorporation for most dopants because they cannot diffuse sufficient rapidly back into the melt[42]. Figure 22 shows a longitudinal cut of a phosphorus-doped crystal with facets and the corresponding resistivity distribution. Facets

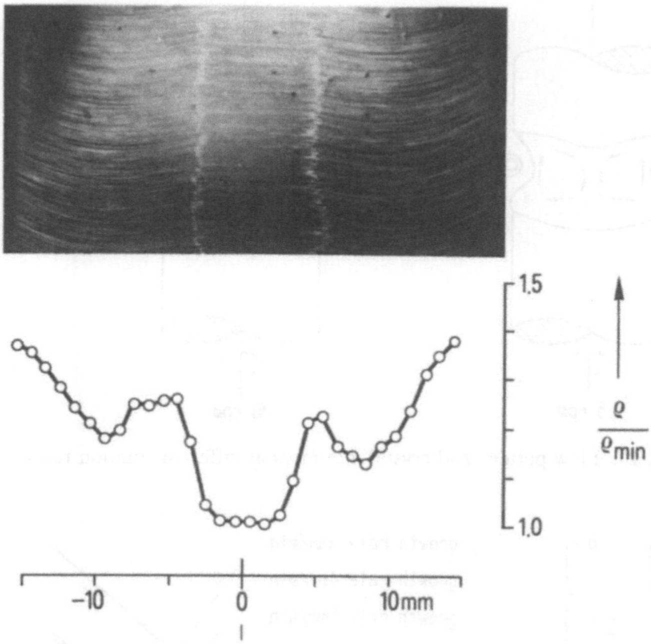

Fig. 22. Longitudinal cut of a [111] crystal with a central facet and the corresponding radial resistivity profile

have not only been observed on convex sections of the growing interface. In some cases, secondary facets are also found in concave portions of the growing interface[43] again with an increased dopant incorporation.

[100] crystals cannot show facet growth because {111} areas do not fit onto the growing interface. Nevertheless, also in this case, a radially inhomogeneous dopant distribution is observed. A possible explanation must take into account the thickness of the dopant diffusion layer forming on the growing interface. Burton et al.[44] have shown that the effective dopant distribution coefficient decreases with diminishing thickness of the diffusion layer, i.e. the dopant concentration in the crystal becomes relatively small if the layer was thin and vice versa. This effect is stronger for group V dopants than for group III dopants. This means that radial dopant variations for group V dopants are larger than for group III dopants. The obviously critical diffusion layer thickness may depend on the flow in the adjacent melt. This flow is caused primarily by the rotation of the growing interface with resulting centrifugal forces. How the shape of the floating zone changes with rotation rate is shown in Fig. 23. Obviously, the originally concave growing interface is transformed into a convex interface with increasing rotation rate. Because of the centrifugal flow distribution, flowless areas of stagnation are formed in the center and at the rim of the growing interface where the diffusion layer seems to remain thicker than in the flow regions. A corresponding dopant distribution is indeed found in phosphorus-doped crystals[1]. Also, the growth rate and of course the crystal diameter play an important role so that still no further explanation of the observed phenomena is possible.

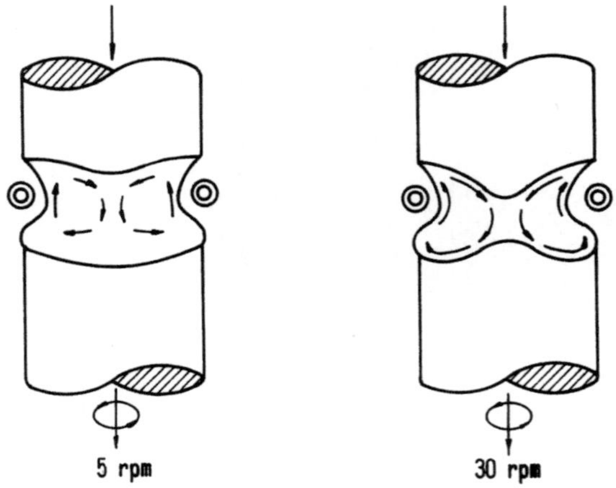

Fig. 23. Flow pattern and crystal interfaces at different rotation rates

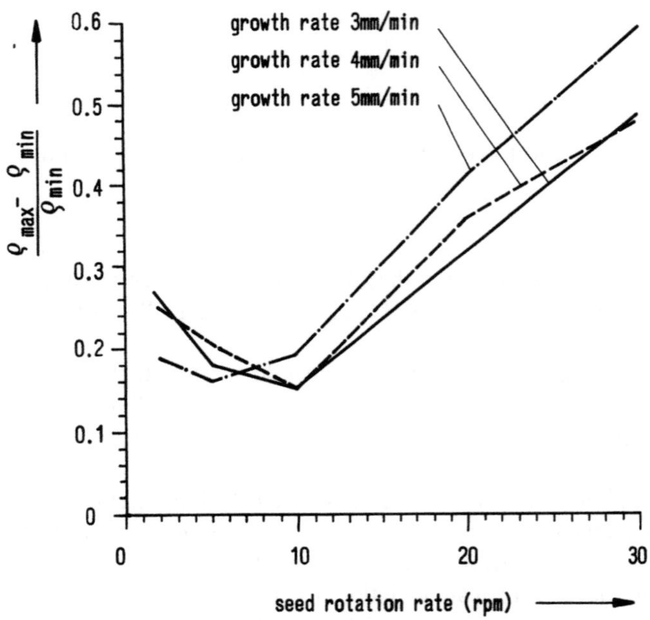

Fig. 24. Radial inhomogeneity of phosphorus-doped [111] crystals grown at different seed rotation rates and growth rates (crystal diameter 34 mm)

Figure 24 shows the interdependence of growth rate, rotation rate and radial dopant distribution for a [111] oriented phosphorus-doped crystal. Note that under optimum conditions a dopand homogeneity of ± 7.5% can be achieved. By counterrotating the feed rod, a small further improvement is possible, but this effect is by far not as pronounced as the seed rotation influence.

In the context of faceted growth, a further effect must be mentioned which occurs in dislocation-free [111] oriented crystals. Here, the cross section does not remain round as in dislocated crystals. Bulges grow out at the crystal periphery in addition to the growth lines. Exactly oriented crystals of a diameter up to 40 mm show three such bulges which lead to a slightly triangular cross section. Larger crystals can contain even more bulges which are not oriented crystallographically. Their occurrence depends on the growth conditions. If the crystal axis is slightly misoriented, i.e. if it deviates, for instance, by 0.7° from the [111] orientation only one bulge forms where the plane of the rotation axis and the [111] direction intersect the cylindrical crystal surface. In order to keep the grinding loss low it is preferable to have only one bulge. Hence, it can be advantageous to grow [111] crystals with some misorientation. Ciszek[45] studied the formation mechanism of these peripheral bulges. He found facets at the periphery of the growing interface where radial outgrowth occurs more readily than axial growth; thus, bulges must form (Fig. 25). If the crystal is misoriented, only such a facet and consequently one bulge can be generated at the position of maximum supercooling (Fig. 25, left). The outgrowth of a bulge is counteracted by the surface tension of the melt so that the bulge cannot grow infinitely.

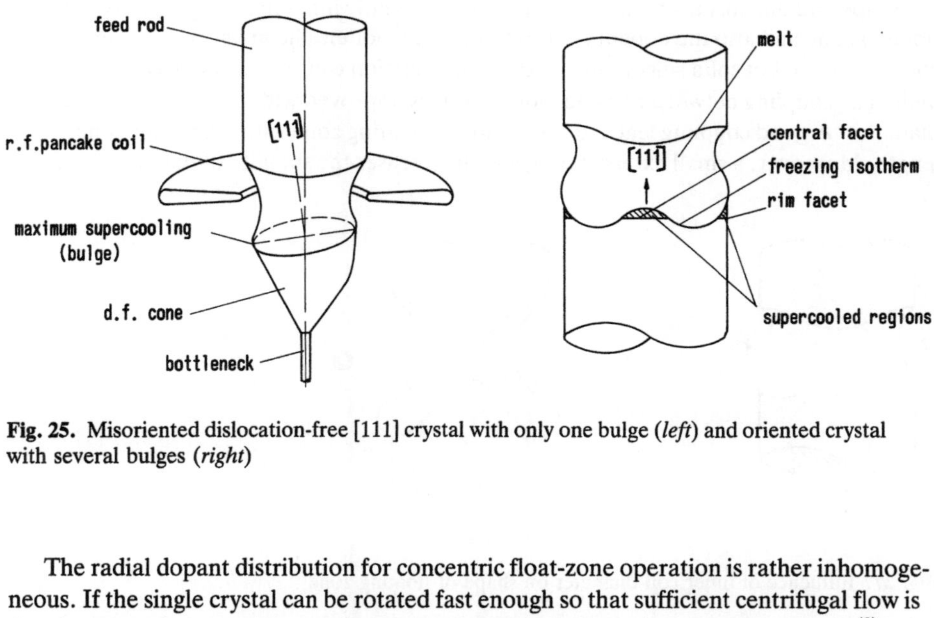

Fig. 25. Misoriented dislocation-free [111] crystal with only one bulge (*left*) and oriented crystal with several bulges (*right*)

The radial dopant distribution for concentric float-zone operation is rather inhomogeneous. If the single crystal can be rotated fast enough so that sufficient centrifugal flow is developed, an improved dopant homogeneity can be achieved by eccentric growth[40]. Here, the growing crystal is in an eccentric position relative to coil and feed rod, the crystal axis being moved to the side parallel to the coil axis by several mm. In Fig. 26 the concentric and eccentric growth modes are compared. In the eccentric mode, the growing interface remains concave even at rapid seed rotation rates. It is a drawback that the eccentric method is not very useful for the growth of large-diameter crystals because here the rotation rate must be kept low to avoid large centrifugal forces which would cause spiral growth or even loss of the melt. The eccentric mode results in improved macroscopic radial homogeneity whereas the microscopic axial homogeneity can be deteriorated. The reason is that the interface temperature variations per rotation are increased because of the asymmetric heater field[1].

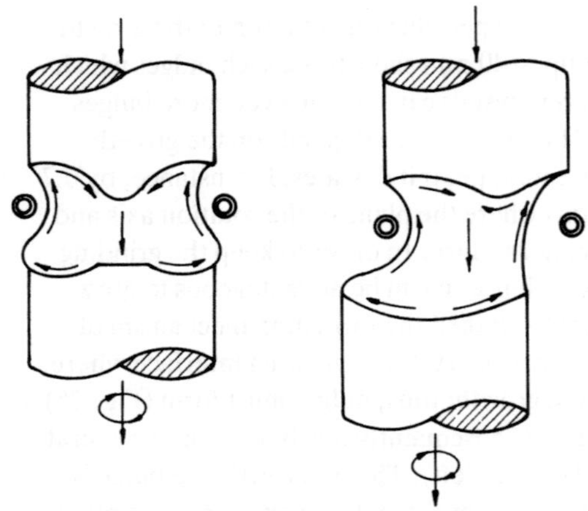

Fig. 26. Flow pattern of concentric and eccentric crystal growth

Shape and dimensions of the r.f. induction coil greatly influence the geometry of the floating zone and also the dopant distribution. Since molten silicon has twenty times the conductivity of hot solid silicon, the field of the induction coil concentrates upon the melt. The coupling between coil and molten zone is improved with decreasing inner coil diameter. A good coupling leads to a well molten floating zone with high inner zone length. Moreover, a small external zone length increases the stability of the molten zone.

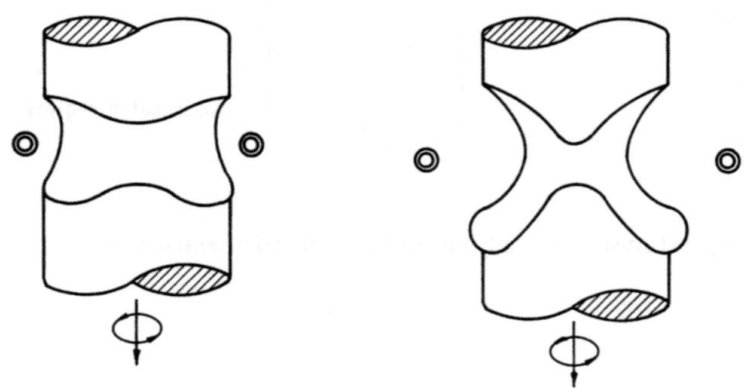

Fig. 27. Influence of inner coil diameter on shape of floating zone

Figure 27 shows the external zone length which is large for coils with large inner diameter because of the considerable stray field. Consequently, the growing interface becomes convex with an unfavorable dopant distribution. It has already been mentioned that the growth rate markedly affects the inner shape of the melt. It defines the heat of fusion per unit area which is used up for melting the feed rod. For stationary growth conditions, the same heat amount is liberated on the growing interface as latent heat of crystallization. The proportion of this latent heat in the total power balance of the system increases with growing diameter. For crystals of 70 mm diameter and a growth rate of 4 mm/min, it is already 1.1 kW. Because of the limited thermal conductivity of the solid portions near the melt, the heat used up on the melting interface and recovered on the growing interface especially influences the inner portions of the interfaces whereas the decisive thermal factor at the rod periphery near the floating zone is radiation. The actual influence of the growth rate on the melt shape at low seed rotation rates is demonstrated in Fig. 28. With increasing growth rate, the melting interface becomes increasingly convex, the growing interface more and more concave. The relative position of the coil shifts more and more toward the melting interface. At fast seed rotation rates, the influence of this rotation clearly dominates the influence of the growth rate. The maximum growth rate obtainable depends strongly on the crystal diameter. Ten mm crystals can be grown at a rate of 15 mm/min whereas it is difficult to grow 80 mm crystals at growth rates higher than 5 mm/min.

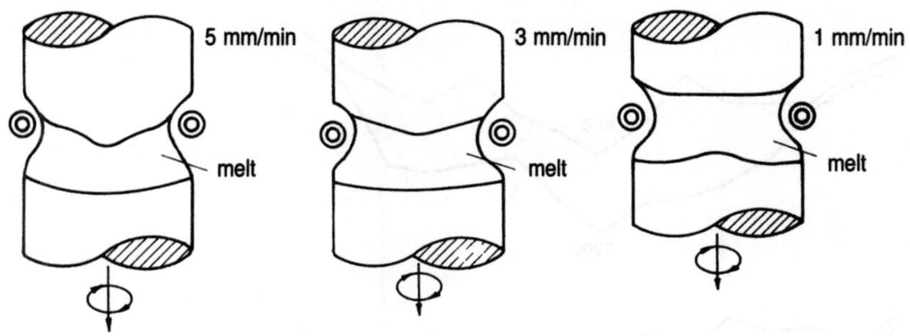

Fig. 28. Influence of growth rate on the shape of the floating zone

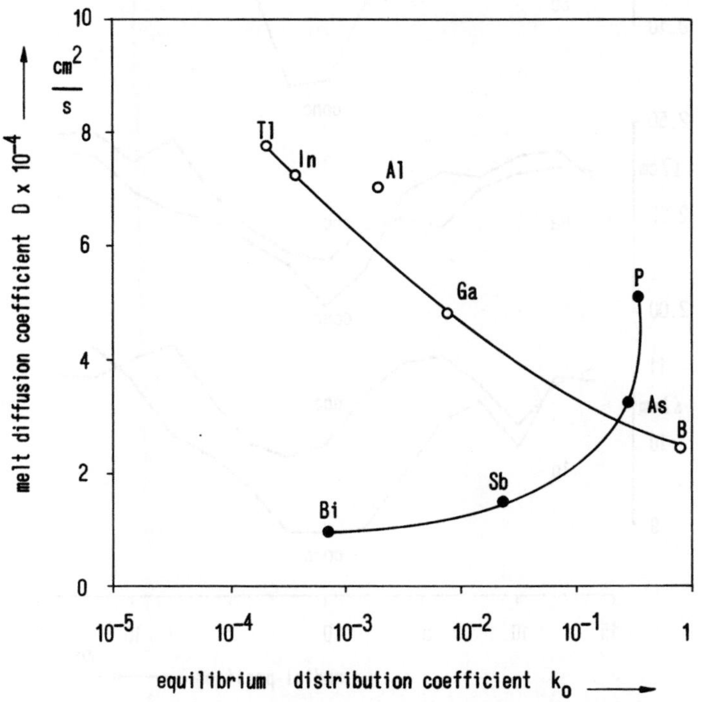

Fig. 29. Equilibrium distribution coefficient and melt diffusion coefficient for important dopants

Group III dopants show a higher dopant homogeneity than group V elements. This is attributed to the different diffusion behaviour and thus different diffusion layer thickness of these elements. The growth rate is again an important parameter (Fig. 10). A comparison of equilibrium distribution coefficients and diffusion coefficients of dopants shows that the k_{eff} values of the group V elements antimony and bismuth vary considerably with rising growth rate because of the small diffusion coefficients of these elements (Fig. 29)[1)]. Figure 30 gives examples for the radial dopant distribution of several dopants. To avoid the influence of faceting on the distribution of the dopant, all crystals were grown in [100] orientation, both in the concentric and eccentric growth method. The homogenizing effect of the eccentric method should be noted[1)].

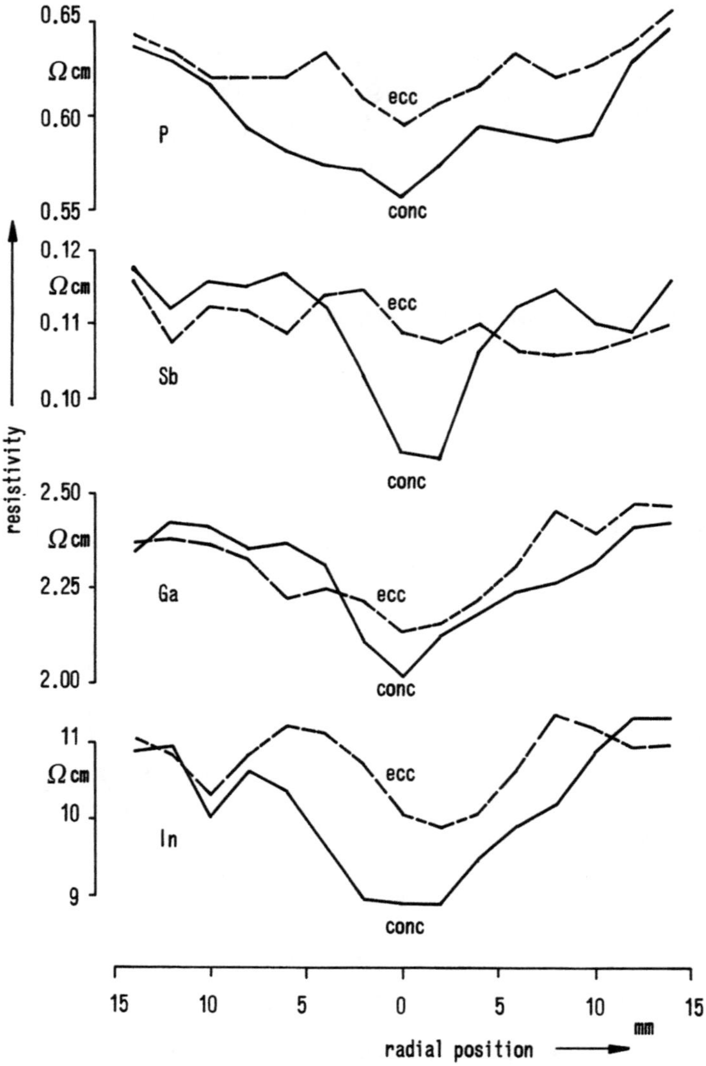

Fig. 30. Experimental radial resistivity profiles for various dopants

The axial dopant profile depends on the distribution coefficient and doping method. Some doping methods are described in Sect. 3.2.6.

3.2.6 Doping Methods

In general, silicon single crystals doped with group III or group V dopants at certain concentration levels are required. The many parameters which influence the radial dopant distribution in the growing crystal have been discussed in Sect. 3.2.5. However, the macroscopic axial distribution must be made as uniform as possible too. Several methods for an axial doping of float-zone crystals are described in this section. It has been shown that dopants can be introduced already during growth of the polycrystal. Three main methods can be distinguished as depicted in Sect. 3.1. The generation method is particularly suited for boron doping where segregation and evaporation losses can be neglected. Other methods must be applied to dopants which segregate and evaporate. Practically no problems arise with axial homogeneity in the case of boron doping because of the high effective segregation coefficient k_{eff} of 0.9 (Fig. 31). A floating zone passed through a uniformly doped feed rod will not markedly change the axial distribution. For phosphorus with k_{eff} 0.5, the inhomogeneity near the seed end of the crystal is already considerably more pronounced and for antimony with k_{eff} 0.07, no equilibrium is reached for actual growth lengths. An additional difficulty with these dopants arises during vacuum growth because their evaporation rate is high. This evaporation rate depends on zone length and growth rate[46, 47].

A doping method for dopants with extremely small distribution coefficients as for instance indium ($k_0 = 4 \times 10^{-4}$) or gallium ($k_0 = 8 \times 10^{-3}$) is pill doping[4]. In this method, the doping pill is inserted into the melt shortly after seeding. Because of the small distribution coefficient of the dopant, which causes an almost total rejection of the dopant at the growing interface into the floating zone, the content of dopant of the melt

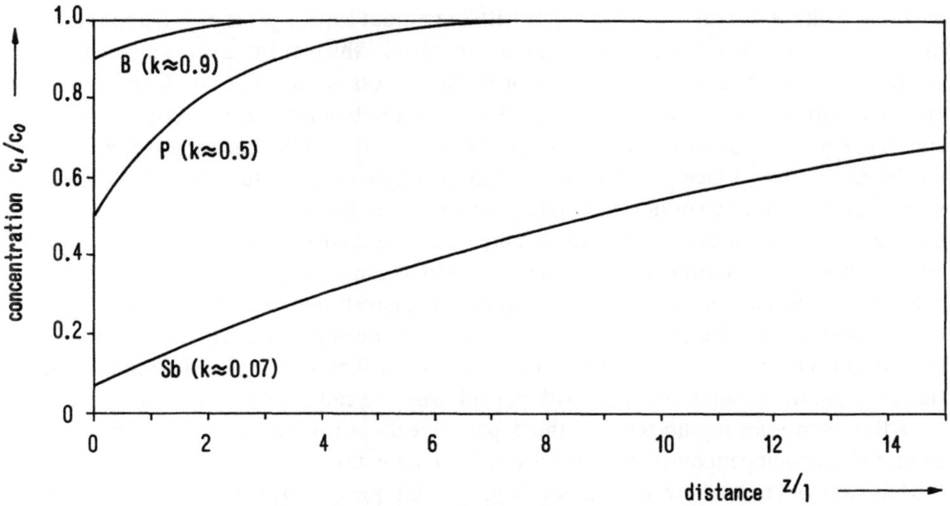

Fig. 31. Calculated axial dopant concentration profiles after one pass

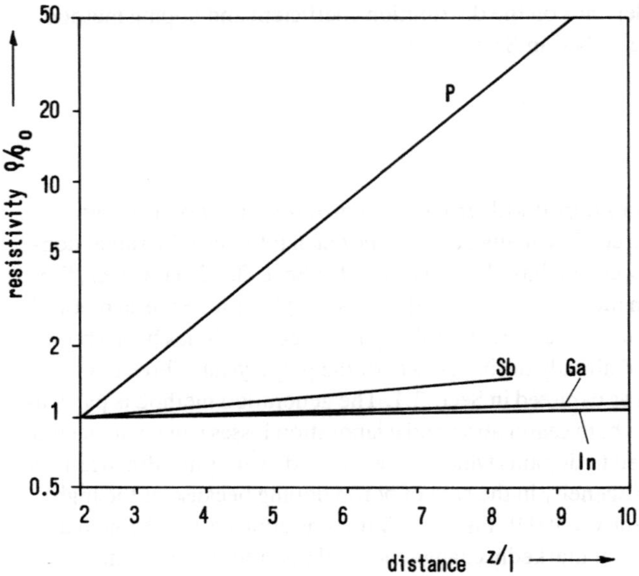

Fig. 32. Experimental axial resistivity profiles of pill-doped crystals

practically does not change up to the growth end. Hence, also the small portion of dopant, which freezes out, remains almost constant (Fig. 32). Of course, evaporation must be prohibited, i.e. the process must take place in a gas atmosphere. The amount of dopant necessary for a certain doping level follows simply from its effective distribution coefficient and the volume of the floating zone. A doping method for dopants with higher distribution coefficients such as phosphorus uses gaseous compounds of the dopant which are blown upon the melt during the float-zone process[1)]. The compound decomposes on the melt surface and only the dopant enters the melt. The method can be adapted both to crystal growth under vacuum or in an inert gas. When performing this process under vacuum, a dilute stream of dopant vapor with or without carrier gas is blown upon the melt. Since only small amounts of dopant are involved, which must be introduced precisely into the melt, an adequate utilization of the process is not easy to perform. The critical parameters must be adjusted experimentally to obtain the desired dopant concentrations, e.g. for phosphorus-doped material from 3×10^{13} to 10^{15} at/cm^3 giving resistivities from 200 to 5 Ωcm. To obtain the desired dopant concentration right from the start of growth, one can either predope by blowing upon the standing melt prior to growth or start the doping cycle with an increased dopant concentration. With gas doping of phosphorus, a precision of the doping level better than ± 10% can be achieved. If the process is performed in an inert gas atmosphere, the gas stream must not necessarily be blown upon the floating zone. The partial pressure of the doping gas can be controlled in the working chamber. This modification of the process allows for higher dopant concentrations than the vacuum process. With phosphorus one can dope down 10 to 0.01 Ωcm.

All conventional methods where the dopant freezes out at the growing interface cannot yield homogeneously doped crystals. Until now, this can be achieved only with the basically old neutron-transmutation doping technique (NTD) which has experienced a dramatic revival recently. The NTD technique was first mentioned in the early fifties[48)].

After improving and adapting it to technical requirements, this doping method has become a very important standard method especially for power-device material[49–51]. In contrast to all the other doping techniques described above, with the NTD process phosphorus doping is possible within dislocation-free, as-grown, undoped single crystals of any diameter. Since phosphorus is the most important starting dopant for power-device production, the device engineer must be very happy that just this dopant can be generated so uniformly within the starting material. The NTD process can be applied because in natural silicon the silicon isotope ^{30}Si occurs with an abundance of about 3%. By irradiation with thermal neutrons this isotope is transformed into ^{31}Si which decays into stable ^{31}P with a half-life time of 3.6 hours

$$^{30}\mathrm{Si}\,(n,\gamma) \longrightarrow {}^{31}\mathrm{Si} \xrightarrow{3{,}6\,h} {}^{31}\mathrm{P} + \beta$$

The process takes place in a nuclear reactor with prevailing thermal neutrons and a homogeneous neutron flux. The number of the ^{31}P atoms generated is proportional to this flux and to the exposition time. A doping precision of better than ± 5%, macroscopic and microscopic, axial and radial is standard. However, the crystal-lattice perfection is strongly disturbed by irridiation, and the phosphorus generated must be transported to lattice sites. This is achieved by annealing of the irradiated crystal at temperatures between 700 and 800 °C. In a side reaction which causes some rapidly decaying radioactivity, small amounts of sulfur are formed. After about 14 days, this radioactivity has decreased to safe levels. The NTD process is mainly useful for relatively low phosphorus concentrations as desirable for the power-device manufacture. High concentrations, i.e. resistivities below 10 Ωcm, must be produced with conventional methods for economic reasons.

Some remarks concerning the measurement of dopant concentration and distribution may be useful now. Since the electric resistivity is inversely proportional to dopant concentration for uncompensated material, the resistivity is the parameter which is important here. It can be measured by means of simple tools which give reliable results. The integral resistivity of the as-grown crystal rod is measured with a low spatial resolution by the two-point probe method (Fig. 33)[52]. A current is fed to the crystal ends. The voltage along the crystal is measured between two contacts which are generally 10 mm

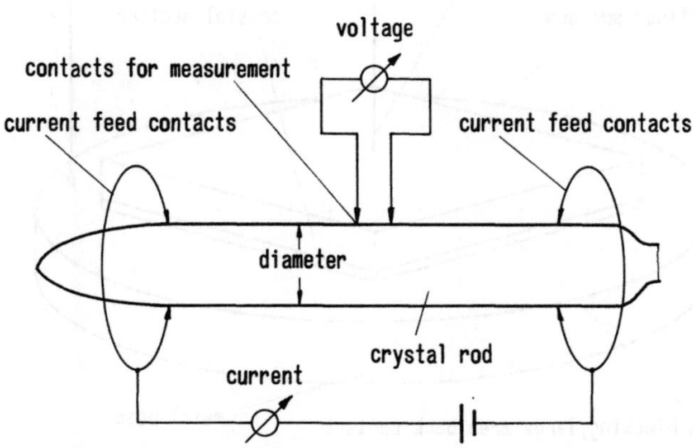

Fig. 33. Two-point probe measurement for axial resistivity determination

apart. Thus, for a crystal diameter of 50 mm, the measurement covers a volume of 20 cm^3. A considerably higher spatial resolution is achieved using the four-point probe method where four equidistant point spring-contacts in a line are applied[53]. It is particularly useful for the measurement of macroscopic radial resistivity distributions on wafers (Fig. 34). Through the two outside contacts, current is introduced into the sample whereas the inner two contacts measure the voltage. The voume measured with this set-up is about 10^{-4} cm^3. An even higher spatial resolution is possible with Mazur's spread-

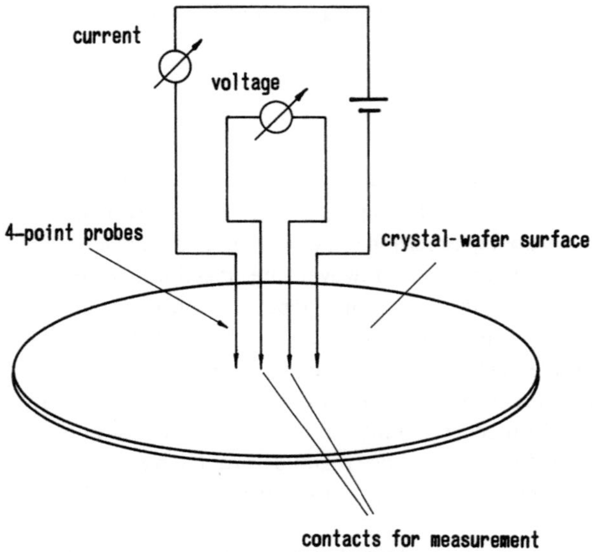

Fig. 34. Four-point probe measurement for radial resistivity determination

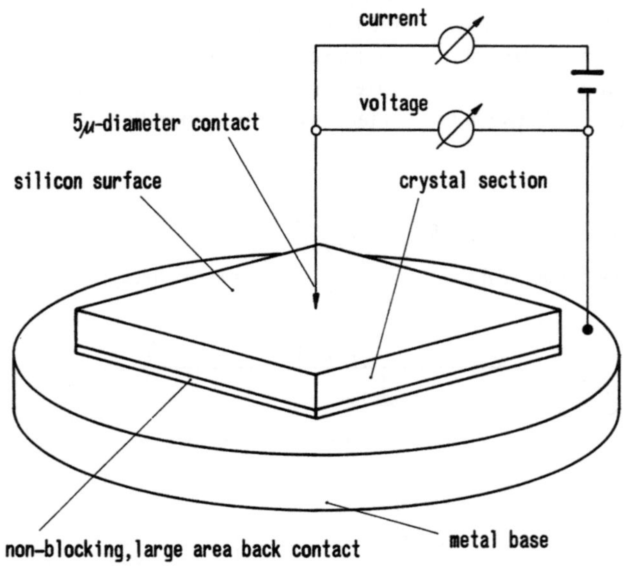

Fig. 35. Spreading-resistance technique

ing-resistance technique[54] where a volume of about 10^{-9} cm^3 can be measured (Fig. 35). Through a fine point contact, a current is fed into the sample which has a large non-blocking back contact. From this current and the voltage between the contacts, the spreading resistance, which is a function of the sample resistivity, can be calculated. The method is especially suited for the measurement of dopant microinhomogeneities which can also be made visible by etching as so-called striations (Fig. 36)[55, 56]. These striations

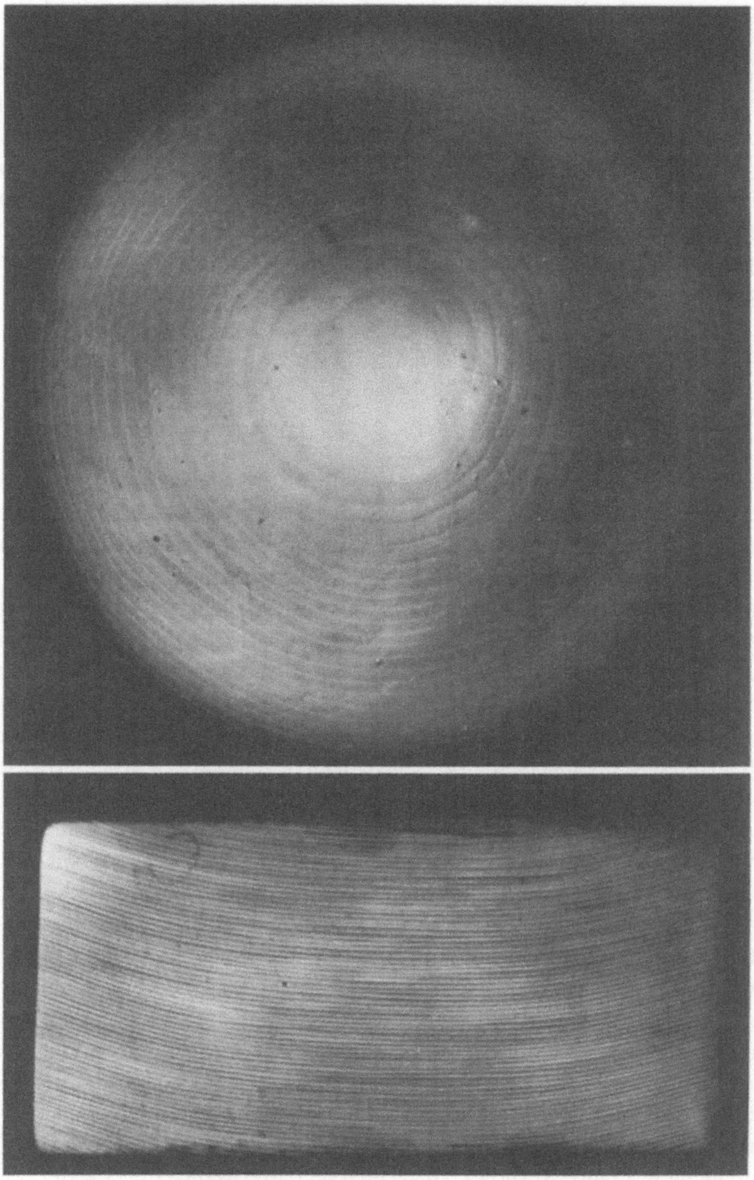

Fig. 36. Cross section and longitudinal cut of etched striation pattern of a float-zone crystal

approximately mark the shape of the growing interface. They are caused by temperature variations near the growing interface which microscopically change the growth rate and dopant distribution. Non-rotational and rotational striations are distinguished. Rotationals repeat periodically and their period is given by the growth rate and seed rotation rate[57]. They are caused by the relatively unsymmetric heater field which induces temperature variations on the rotated growing interface. The resulting resistivity changes are between ± 10 and ± 50%. After etching, non-rotationals generally show weaker contours than rotationals which indicates a lower dopant variation. They are not periodical and have a smaller distance[58]. Non-rotational striations can have quite different sources, e.g. vibrations of the melt or temperature oscillations in the melt. A periodic supercooling of the melt near the growing interface has been completely ruled out experimentally.

Finally, another method for the quantitative determination of impurities must be mentioned, the infrared method[59]. In the large ranges of the infrared spectrum above a wavelength of 1.1 μm, the silicon lattice has no absorption. Dopants and impurities on lattice sites give rise to absorption peaks at well-defined wavelengths, e.g. boron at 31.3 μm and phosphorus at 31.6 μm. Polished crystal wafers are used in the method, oxygen and carbon can be measured at room temperature whereas phosphorus and boron require liquid helium cooling for a sensitive measurement.

3.2.7 Crystal Defects

The history of silicon crystal growth is also the history of silicon crystal defects. Quite often, avoiding one kind of defect by improving the growth technique leads to the detection or even formation of new, mostly smaller defects. No end of this interdependence between growth technique and defect sophistication is in sight and also the requirements of the device manufacturers become more stringent. Crystal defects can be distiguished from two-dimensional stacking faults according to their geometric dimensions, e.g. twins, down to one-dimensional dislocations and zero-dimensional point defects. The latter often aggregate to clusters; thus, they must rather be regarded as three-dimensional. The transitions can be ambiguous. The oldest method for detecting a defect is by etching. Many special methods have been developed[35]. Particularly useful for the detection of several kinds of defects is Sirtl's method[56] which, for instance, reveals dislocations with a clear contrast, especially on {111} planes. At the site where a dislocation intersects the crystal surface, a deep more or less triangular etch pit is revealed. Dislocations can be made visible non-destructively by X-ray topography. Thin planar arrays can be detected in thin foils by transmission electron microscopy. An old detection method for decorated dislocations in larger samples is the infrared transmission topography[60]. Edge and screw dislocations are distinguished. However, in silicon crystals, mostly mixed dislocations with components of both dislocation types are observed. Dislocations can aggregate to form grain boundaries. According to the crystallographic orientation, slippage and lineage boundaries are distinguished. Whereas slippage can be generated in the hot solid crystal by stress application and can partly be annealed away by a stress-free temperature treatment, lineage has so far only been observed after solification from the melt. Figure 37 shows typical etch pit patterns on a {111} wafer. Grain boundaries, especially lineage and high etch pit density impair device performance.

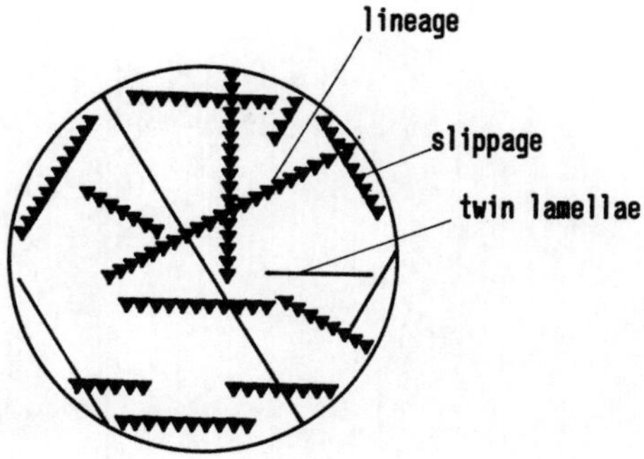

Fig. 37. Twin lamellae lineage and slippage on {111} wafer

In dislocated crystals, the etch pit density increases with increasing diameter because the thermomechanical stress rises with the diameter. Conditions for minimum stress growth require either small temperature gradients or a three-dimensional uniform temperature gradient distribution. Both conditions can be approximated only very roughly during actual growth processes. A concave growing interface can be advantageous, not only because it imports more stability to the melt but also because the critical transition to the concave isotherms in the cooling crystal can be more stress-free than with a flat or even convex interface. Crystals with diameters above 45 mm can be grown more easily dislocation-free than dislocated, because in a dislocated crystal it is difficult to avoid extensive dislocation multiplication under the existing thermomechanical stress. On the other hand, this stress may not be sufficiently large to initiate formation of dislocations. Dislocated single crystals always show residual stress whereas dislocation-free crystals seem to be at least macroscopically stress-free. This can be understood as a consequence of the purely elastic behavior of dislocation-free material under thermomechanical stress.

For instance, experiments have shown that dislocation-free stressed crystals are fully elastic at 800 °C whereas dislocated material already exhibits plastic deformation. A consequence are effects which can be observed in originally dislocation-free crystals of large diameters which suddenly develop dislocations, e.g. at the crystal end. During the hot phase, the dislocation-free region is stressed plastically by forming dislocations which reduce stress. On cooling, both regions fit increasingly less. Finally, they are separated by cracking.

The same etching technique which reveals dislocations also reveals swirl defects. Swirls were first observed in 1965[61)]. Figure 38 shows etched-out swirl defects on a dislocation-free silicon wafer. Under the microscope, swirls can be easily distinguished from striations, which also form spiral patterns on wafers, but are lines and not points. Swirls consist of rows of etch pits which are flat in contrast to the deep dislocation etch pits. A single swirl etch pit disappears on etching, new ones form at new spots, indicating the point defect character of swirls. Dislocation etch pits generally retain their position on etching. De Kock[62)] has distinguished large A-defects from small B-defects, the B-defects aggregating to form A-defects. The nature of A-defects has been clarified by high-voltage electron microscopy[63–65)]. They consist of collapsed interstitial point defects which generate small intrinsic dislocation loops. It is much more difficult to clarify the nature of B-defects which show no contrast under the high-voltage electron microscope. It is assumed that interstitials first form small clusters which are B-defects. Impurities, especially carbon, seem to play a dominant role here. High carbon concentrations finally

Fig. 38. Swirl pattern on wafer

lead to the formation of large A-defects and stacking faults which strongly deteriorate the device performance[66]. Only under sufficiently low growth rates have the B-defects sufficient time to form A-defects or even macroscopic dislocations. So A-defect formation can be suppressed by adequately rapid crystal growth[62].

Float-zone crystal growth in the presence of hydrogen can lead to very strong defects[1]. They can be revealed by etching of wafers or cross sections of a crystal with acids or hydroxides (Fig. 39). Deep holes, grooves or craters are formed in the inner regions of the crystal. A hydroxide etch leads to negatives of crystals, i.e. crystallographi-

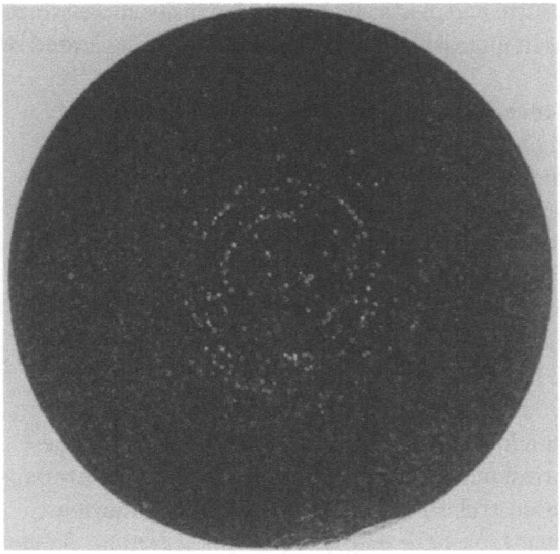

Fig. 39. Hydrogen defects on hydroxide-etched wafer

cally well oriented craters. How the defect is caused is not yet clear, possibly by hydrogen clustering with interstitials. Devices made from crystals with hydrogen defects have no blocking capability. Today, hydrogen is no longer used as a growth environment, only argon is employed for dislocation-free growth.

In float-zone crystals of diameters above 45 mm another defect has been found. After polish etching of material containing the defect, marked depressions are found in the crystal interior[67]. On wafers, they often have the shape of a frying pan (Fig. 40).

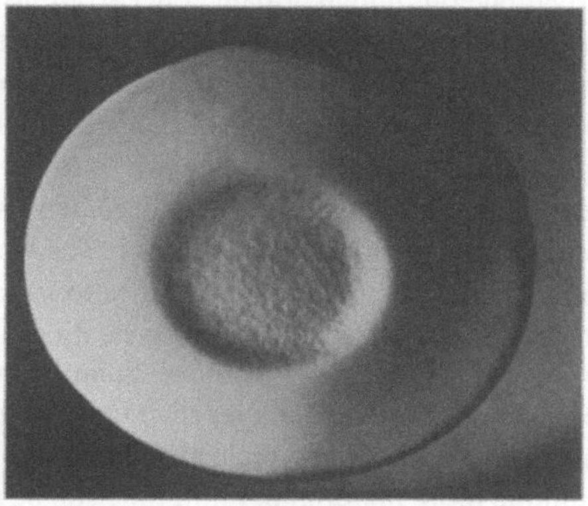

Fig. 40. Etching depression on a polish-etched, dislocation-free wafer

Although looking serious, these etching depressions have not yet shown an influence on device performance. Of course, crystal growth successfully tries to avoid the formation of defects. A possible cause for etching depressions is the thermomechanical stress which occurs in the cooling phase after freezing. Probably, the presence of point defects is again important. Since the crystal interior is in a compressed state because it is less rapidly cooled than the periphery, the concentration of interstitials in the inner bulk may be larger than that at the outside. Gold decoration experiments in crystals with etching depression show a decoration pattern which closely resembles the depression geometry. Since the gold distribution is similar to that in material without depression, although with less contrast, one must conclude that the defect is basically present in any large-diameter dislocation-free material. The formation of a depression then depends only on the amount of defects.

4 Conclusions

For a summary of the essential properties of the float-zone method, a comparison with the Czochralski technique can be helpful. What are the advantages and drawbacks of both methods? As far as purity is concerned, the float-zone method is, without doubt, far superior to the Czochralski technique. Because of the contact of the melt with the crucible, uncompensated resistivities above 20 Ωcm are hardly obtainable whereas the float-zone process easily yields resistivities above 100 Ωcm, even in one pass. The use of hyperpure polycrystalline feed rods is to be recommended only for float-zone growth because their purity does not change. Moreover, by applying multi-pass zone melting, the final purity can be enhanced considerably. Resistivities close to intrinsic have been reached with this method[68]. Float-zone crystals are pure not only with respect to dopants but also as far as nondoping impurities such as carbon, oxygen, heavy metals, and others are concerned. The superior neutron transmutation doping technique for medium-to-weak phosphorus concentrations requires a starting material of high purity. Thus, for this method only undoped float-zone material is useful. In contrast, the Czochralski method is advantageous if high dopant concentrations of any dopant are required as for instance for substrate material for epitaxy. Such a material can hardly be produced by the float-zone method. The growth of large-diameter crystals with suitable structure is certainly more easily achieved with the Czochralski method than with the float-zone method. The largest diameter which can be grown by latter method at present is about 100 mm. Much larger diameters can hardly be realized because of the stability problems arising from the molten zone under gravity conditions. The heating power supply problems are relatively simple in the Czochralski growth since resistance heaters can be used here whereas the floating zone requires r.f. induction heating. In general, the thermal growth conditions are less critical with the crucible technique as the temperature gradients in the melt and in the growing crystal are smaller than for float-zone growth. On the other hand, as the growth rate in the float-zone method is about three times larger than for the Czochralski technique more material per unit time can be grown with this method at the same diameter. The lack of an expensive crucible is another advantage of the float-zone method. Improved Czochralski techniques thus allow for the growth of several crystals from one crucible. A very important presupposition for effective float-zone growth is a good quality of the polycrystalline feed material, not only as far as its purity is concerned but also with respect to its grain size and surface structure. Hence, progress in the floating zone technique is closely coupled with progress of the polycrystal deposition process. Regarding the power system, the Czochralski technique has the advantage of needing no r.f. generator. However, the long growth cycles and large crucibles involved at least partly compensate the power yield advantage of the simple Czochralski power system. For semiconductor-grade silicon, the power efficiency of growth systems at present does not play an important role. This will certainly change for solar-grade material. The demand for cheap silicon already gives rise to the development of new growth techniques which may also lead to new ideas and techniques in the field of classical silicon crystal growth.

5 References

1. Keller, W., Mühlbauer, A.: Floating-Zone Silicon, Marcel Dekker, New York, 1981
2. Czochralski, J.: Z. Phys. Chem. *92,* 219 (1918)
3. Teal, G. T., Little, J. B.: Phys. Rev. *78,* 657 (1950)
4. Pfann, W. G.: Trans AIME *194,* 747 (1952) and: Zone Melting, 2nd edit., Wiley, New York, 1966
5. Theuerer, H. C.: USP 3 060 123 filed Dec. *17,* 1952, patented Oct. 23, 1962
6. Keck, P. H., Golay, M. J. E.: Phys. Rev. *89,* 1297 (1953)
7. Emeis, R.: Z. Naturforsch. *9 a,* 67 (1954)

8. Müller, S.: Z. Naturforsch. *9b,* 504 (1954)
9. Keck, P. H. et al.: Rev. Sci. Instr. *25,* 331 (1954)
10. Emeis, R., Henker, H.: DBP 975158 filed Dec. 31, 1953 patented Sept. 14, 1961
11. Schweickert, H., Gutsche, H., Emeis, R.: DBP 1177119, filed Nov. 2, 1955, patented May 20, 1965; USP 3030189, filed May 19, 1958, patented Apr. 17, 1962
12. Bischoff, F.: DBP 1102117, 1140549, filed May 18, 1954 USP 3146123
13. Schweickert, H., Reuschel, K., Gutsche, H.: DBP 1061593
14. Van Arkel, A. E., de Boer, J. H.: Z. Anorg. Allg. Chem. *148,* 345 (1925)
15. Hölbling, R.: Z. Angew. Chem. *40,* 655 (1927)
16. Amberger, E.: Doctoral Thesis, Univ. Munich, 1956
17. Dietze, W.: Metall *34,* 676 (1980)
18. Sirtl, E., Reuschel, K.: Z. Anorgn. Allg. Chem. *332,* 113 (1964)
19. Hunt, L. P., Sirtl, E.: J. Electrochem. Soc. *119,* 1741 (1972)
    Sirtl, E., Hunt, L. P., Sawyer, D. H.: J. Electrochem. Soc. *121,* 919 (1974)
20. Sirtl, E.: DBP 1193918, filed June 14, 1960, patented June 22, 1967, USP 3172857
21. Hoffmann, A. et al.: DBP 1153540, filed Sept. 10, 1958, patented March 12, 1964, USP 2970111
22. Brice, J. C.: The Growth of Crystals from Liquids, p. 268, North-Holland/American Elsevier, New York, 1973
23. Keller, W.: J. Cryst. Growth *36,* 215 (1976)
24. Chang, C. E., Wilcox, W. R.: J. Cryst. Growth *28,* 8 (1975)
25. Keller, W.: DBP 2358300, filed Nov. 22, 1973, patented July 20, 1978; USP 3923468, filed Nov. 20, 1974, patented Dec. 2, 1975; USP 3961906, filed May 27, 1975, patented June 8, 1976
26. Herzer, H., Zauhar, H.: USP 4060392, filed May 5, 1976, patented Nov. 29, 1977
27. Gupta, K. P., Gregory, O., Rossnick, M.: J. Cryst. Growth *44,* 526 (1978)
28. Keller, W.: DBP 2160694, filed Dec. 7, 1971, patented Febr. 2, 1980; USP 3827017, filed Nov. 24, 1972, patented July 30, 1974
29. Keller, W.: DBP 1913881, filed March 19, 1969, patented May 27, 1971; USP 3644151, filed March 13, 1970, patented Feb. 22, 1972
30. Keller, W.: DBP 1148525, filed May 29, 1959, patented Dec. 27, 1963
31. Crosby, L. R., Stewart, H. M.: USP 3249406, filed Jan. 8, 1963, patented May 3, 1966
32. Emeis, R.: Internal Siemens Report 112, Pretzfeld, 1954
33. Rummel, T., Keller, W., Quast, H. F.: DBP 1153908, filed April 22, 1958, patented April 9, 1964; USP 2913561 filed April 13, 1959, patented Nov. 17, 1959
34. DIN 50433, part 2 (German standard for optical orientation), Berlin 1976
35. Bogenschütz, A. F.: Ätzpraxis für Halbleiter, Hanser, Munich, 1967
36. Dash, W. C.: Growth and Perfection of Crystals, p. 361, Doremus, R. H., Roberts, B. W., Turnbull, D. (Eds.), Wiley, New York, 1958
37. Ziegler, G.: Z. Naturforsch. *16a,* 219 (1961)
38. Keller, W.: DBP 1094710, filed Febr. 19, 1958, patented Febr. 20, 1969; USP 3159459, filed Febr. 18, 1959, patented Dec. 1, 1964
39. Keller, W., Kramer, H.: DBP 1719021, filed July 13, 1963, patented April 23, 1970; USP 3563810, filed July 10, 1964, patented Febr. 16, 1971
40. Keller, W.: DBP 1218404, filed Febr. 1, 1964, patented Dec. 29, 1966; USP 3414388, filed Jan. 29, 1965, patented Dec. 3, 1968
41. Ciszek, T. F. in: Semiconductor Silicon, Kern, E. L., Haberecht, R. R. (Eds) p. 156. The Electrochemical Society, Princeton N. J., 1969
42. Trainor, A., Bartlett, B. E.: Solid-State Electron. *2,* 106 (1961)
43. Mühlbauer, A. in: Semiconductor Silicon. Huff, H. R., Burgess, R. R. (Eds.), p. 107. The Electrochemical Society, Princeton N. J., 1973
44. Burton, J. A., Prim, R. C., Slichter, W. P.: J. Chem. Phys. *21,* 1987 (1953)
45. Ciszek, T. F.: J. Cryst. Growth *10,* 263 (1971)
46. Hoffmann, A.: Halbleiterprobleme, vol. 6, p. 152. Vieweg, Braunschweig, 1961
47. Ziegler, G.: Z. Metallk. *49,* 491 (1958)
48. Lark-Horovitz, K.: Proc., Conf., Univers., Reading, Butterworth, London, 1951
49. Schnöller, M. S.: IEEE Trans. Electron Devices ED-*21,* 313 (1974)
50. Herrmann, H. A., Herzer, H.: J. Electrochem. Soc. *122,* 1568 (1975)

51. Janus, H. M., Malmros, O.: IEEE Trans. Electron. Devices ED-*23,* 797 (1976)
52. DIN 50430 (German standard for two-point probe measurements), Berlin, 1971
53. Valdes, L. B.: Proc. I.R.E. *42,* 420 (1954)
54. Mazur, R. G.: J- Electrochem. Soc. *114,* 255 (1967)
55. Mayer, K. R.: J. Electrochem. Soc. *120*, 1780 (1973)
56. Sirtl, E., Adler, A.: Z. Metallk. *52,* 529 (1961)
57. Barthel, J., Jurisch, M.: Kristall und Technik *8,* 199 (1973)
58. Witt, A. F., Gatos, H. C.: J. Electrochem. Soc. *113,* 808 (1966)
59. Kolbesen, B. O.: Appl. Phys. Lett. *27,* 353 (1975)
60. Dash, W. C.: J. Appl. Phys. *27,* 1193 (1956)
61. Plaskett, T. S.: Trans. AIME *233,* 809 (1965)
62. De Kock, A. J. R.: Philips Res. Rep. Suppl. *1* (1973)
63. Kolbesen, B. O., Mayer, K. R., Schuh, G. E.: J. Phys. E *8,* 197 (1975)
64. Föll, H., Kolbesen, B. O.: Appl. Phys. *8,* 319 (1975)
65. Petroff, P. M., de Kock, A. J. R.: J. Cryst. Growth *30,*117 (1975)
66. Kolbesen, B. O., Mühlbauer, A.: Solid-State Electron, to be published
67. Keller, W., Mühlbauer, A.: Inst. Phys. Conf. Ser No. *23,* 538 (1975)
68. Hoffmann, A., Reuschel, K., Rupprecht, H.: J. Phys. Chem. Solids *11,* 284 (1959)

Survey Articles:

W. R. Runyan: Silicon Semiconductor Technology, McGraw-Hill, New York, 1965
J. A. Lenard, E. J. Patzner: SCP Solid-State Technol. 35 (1966)
K. Reuschel: Siemens-Z. *41,* 669 (1967)
H. Muraoka et al.: Toshiba Rev. *40*, 44 (1969)
R. C. Bracken: Electro-Technology, p. 43 (1969)
E. Spenke. in: Semiconductor Silicon, E. L. Kern, R. R. Haberecht (Eds-), p. 1. The Electrochemical Society, Princeton N. J., New York, 1969
W. Keller, H. Stut: Feinwerktechnik *75,* 207 (1971)
W. Dietze, A. Mühlbauer: Chemiker-Ztg. *97,* 151 (1973)
W. Keller: Feinwerktechnik. Micron *77,* 107 (1973)
R. E. Lorenzini, F. S. Neff, D. J. Blair: Solid-State Technol. p. 33, 1974
J. Lenzing: Solid-State Technol. 34 (1975)
H. Herrmann, H. Herzer, E. Sirtl, in: Festkörperprobleme, vol. 15, p. 279. H. J. Queisser (ed.), Pergamon/Vieweg, Braunschweig, 1975
E. Sirtl, in: Semiconductor Silicon, H. R. Huff, E. Sirtl (ed.), p. 4. The Electrochemical Society, Princeton N. J., Philadelphia, 1977
S. E. Bradshaw, J. Goorissen: J. Cryst. Growth *48,* 514 (1980)
B. O. Kolbesen, H. Strunk: Invited paper, ESSDERC, York, G. B. 1980, to be published in Inst. Phys. Conf. Series
E. Spenke, W. Heywang: Phys. Status Solidi A, *64*, 11 (1981)

"Solar" Silicon

Josef Dietl, Dieter Helmreich, and Erhard Sirtl

Heliotronic GmbH, D-8263 Burghausen

A survey is given on the present status of technical development in the field of silicon processing for terrestrial solar cell application. After sketching the long-range targets of predominantly economical character, the different worldwide approaches in both chemical preparation and special crystallization techniques are discussed. In the material sector currently investigated methods to produce silicon are presented together with the basic means of refining and the problem of "solar-grade" quality.

The treatment of unconventional crystallization is split up into ingot (plus slicing) techniques, direct ribbon growth, and methods for film deposition. Finally, material characterization is shown in the light of specific, diagnostic needs in the photovoltaic sector with emphasis being put on the different features of multicrystalline silicon.

Crystals: Growth, Properties and Applications 5
© Springer-Verlag Berlin Heidelberg 1981

List of Abbreviations

PV technology	photovoltaic technology
PV CE	photovoltaic conversion efficiency
AM 0	air mass 0
AM 1	air mass 1
AR coating	anti reflection coating
HEC	high efficiency cell
BSF	back surface field
BSR	back surface reflection
ID saw	internal diameter saw
SAM	scanning Auger microscopy

1 Introduction

1.1 The Role of Silicon as a Semiconductor

Silicon is unchallenged as a semiconductor base material in our present electronics industry. The reasons why it qualifies so strongly for this particular purpose are manyfold. The attractive combination of physical (electrical) properties of silicon and the unique properties of its native oxide layer have been the original factors for its breathtaking evolution in device technology. The majority of reasons, however, for its present status are correlated with industrial prosessing in terms of charge units (economy), reliability (reproducibility), and flexibility, but also its availability.

The latter point, in particular, plays an important role in the different long-term projects on the terrestrial application of solar cells. Practically inexhaustive resources of silicon dioxide form a sound basis even for the most pretentious programs on future alternatives to relieve the present situation in electrical power generation by photovoltaics. Assuming a maximum percentage of 10% to be replaced by the year 2000 would roughly mean a cumulative annual production of 2 million metric tons of crude silicon (based on present solar cell standards)[1].

To illustrate the orders of magnitude that have to be discussed in pertinent programs: Today, the industrial silicon capacity of non-communistic countries (including ferrosilicon and other alloys by their relative Si-content) amounts to some 2 million tons per year. A quarter of it consists of "silicon metal", i.e. metallurgical-grade silicon (MG-Si) of 98% purity, as an average quality[2]. In Fig. 1 the relation of this production capacity is shown in comparison to other materials of relevance[3]. Most of the total silicon quantity is presently used as deoxidizer in the steel industry, the major amount of MG-Si as an alloy

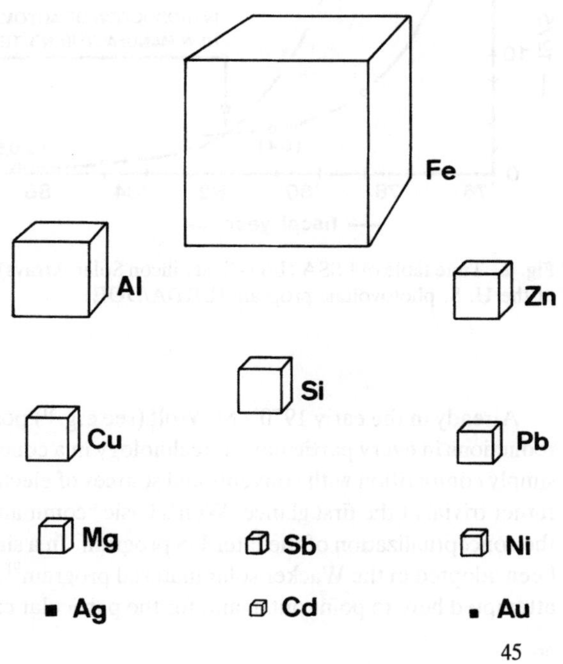

Fig. 1. Relative volumes of worldwide production of different basic materials (elements)

component in the aluminium industry. In 1981, 20% of the MG-Si capacity referred to will be utilized for the preparation of organosilanes (silicone polymers), another 4% – i.e. 1% of the total production of silicon – for manufacturing hyperpure "electronic-grade" silicon (EG-Si).

1.2 Photovoltaic Programs

To initiate the necessary development on a broad basis, photovoltaic programs have been started between 1974 and 1975 in different parts of the world[4–6]. The best known by far is the ERDA/DOE program of the USA conceptualized for the time period 1975–1985. Its essential part in photovoltaics is concentrated in the so-called LSSA Project (see Fig. 2)[7]. The basic idea behind is demonstrated by the targets in price development for solar cell-based modules in peak Watt ($\$/W_p$). The time table is characterized by two different functional systems (Fig. 2) where it can be seen that in the earlier years accomplishments can be expected only by semiconventional approaches on the basis of EG-Si material and space-type solar generator technology. Simultaneously started projects to explore unconventional techniques are planned to form the basis for mass production of solar generators towards the end of the present decade.

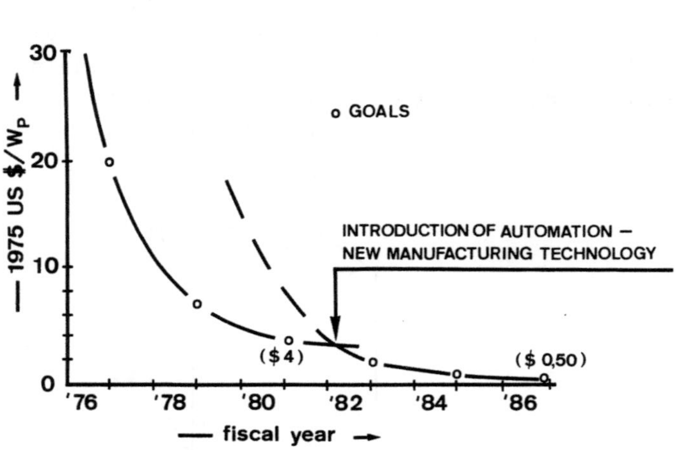

Fig. 2. Time table of LSSA (Low-Cost Silicon Solar Arrays) Project on generator costs – as a part of the U. S. photovoltaic program (ERDA/DOE)

Already in the early 1970's M. Wolf (see e.g.[8]) postulated an approach involving cost reductions in every particular subtechnology in a concerted manner. Its final aim was simply competition with conventional sources of electrical energy. Although looking rather trivial at the first glance, Wolf's basic "commandments" had a strong impact on the conceptualization of the later US program. In a similar way, his general ideas had been adopted in the Wacker solar material program[9] as it is illustrated in Fig. 3. It is attempted here to point out again, for the particular case of material preparation, the

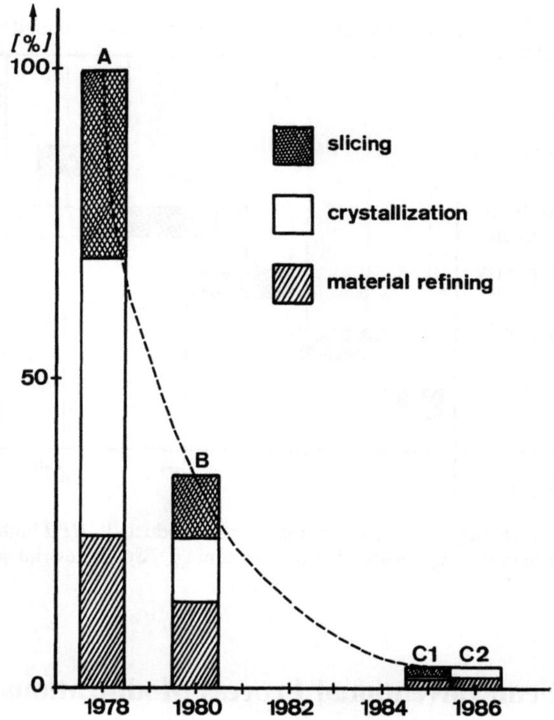

Fig. 3. Planned development of relative costs in the Wacker program on "solar-grade" silicon. Product: Square-shaped slices. *A* Monocrystalline, conventional production; *B* Multicrystalline, pilot production, technology mix; *C* Multicrystalline, feasibility studies (pilot), "all-solar-grade" technology. (*C1* sliced ingots, *C2* ribbons or foils)

necessity of a more or less concurrent and balanced cost reduction for all three technical disciplines, material refining, crystallization and slicing.

1.3 "Solar-Grade" Definition

Under this term a project has been pursued in the framework of the DOE program in the USA aimed at categorizing all elements of relevance with respect to their influence on solar cell degradation[10] (see Sect. 6.1.1). In the meantime, the expression "solar grade" has become a "trade mark" for different products and often did not sufficiently distinguish between specialities in a chemical, crystallographic, or device-related sense.

Primarily, we would like to see this term being confined to the chemical properties of starting material qualities under the special aspects of low-cost terrestrial solar cell technology – and to name it more precisely "terrestrial solar grade" (TSG). Further information on silicon quality ranges and their terminology is given in Fig. 4. Due to a fairly strong correlation with the special state of crystallization and the subsequent treatment during device fabrication, future specifications may have to be worked out for every particular processing sequence – even on the basis of a "standard" indiffused p–n junction. At the time being, the different types of solar cells and their yet unknown variation during the further stages of development make the definition of TSG-Si a rather difficult one. The solar material suppliers, undoubtedly, have to be prepared for a rather wide spectrum of specifications in the future photovoltaics business.

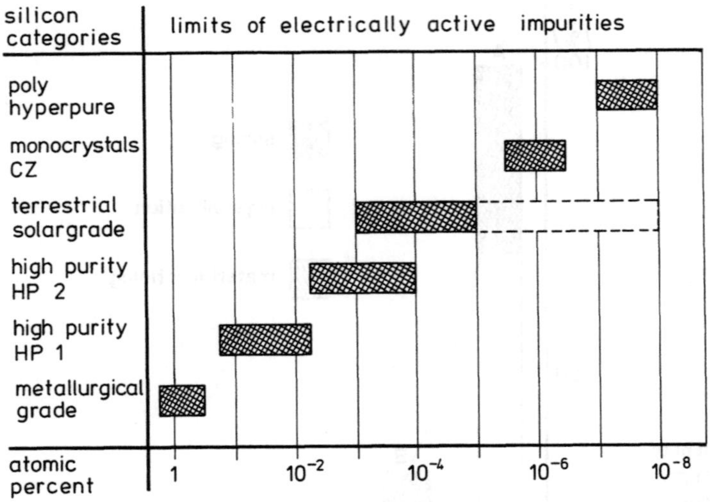

Fig. 4. Quality ranges for different silicon products[11]. *HP 1* high purity, basic stage of refining; *HP 2* high purity, advanced stage of refining; *TSG* terrestrial solar grade, final stage of refining

2 Semiconventional Process Modifications

The final target of the solar program can be reached only by applying unconventional methods in every particular subtechnology. Semiconventional approaches – in base material production, crystallization and in slicing techniques as well – may help to ease the solar material market situation for several more years to come.

For the transitional phase of providing appropriate starting material for the different crystallization facilities, one way will be to live with the present electronic-grade quality of "CVD-Poly". Improvements of the existing technology based on trichlorosilane by modification of the deposition reactor gives promise of more material for solar application at lower cost[12]. The general approach is characterized by abandoning rod growth in favor of larger surface areas of silicon per reactor volume unit. The technical solution offered here is based on fluidized-bed rotary drum or systems[13, 14]. Another route is to transfer the trichlorosilane to dichlorosilane[15]. Dichlorosilane is generated here by catalyzed disproportionation of trichlorosilane.

Tetrachlorosilane, as a by-product, can be recycled to trichlorosilane via hydrogenation (see also Sect. 3.1). Dichlorosilane decomposes more readily than trichlorosilane, resulting in substantially higher deposition rates[16, 17]. Physical state and purity of the silicon product are claimed to be consistent with material requirements for use in the manufacture of high-efficiency solar cells.

Improvements of the existing crystallization techniques in terms of lower cost are restricted to the well-studied Czochralski (CZ) method. Limited dissipation of the heat of crystallization and a critical growth rate in maintaining the monocrystalline state are basic obstacles to a substantial increase in throughput. Analysis of the cost-relevant factors of the CZ process shows that about half of the add-on cost depends on the quartz crucible[18]. Optimum utilization of quartz crucibles can be achieved by continuous melt replenishment[19–21] thus eliminating the need for subjecting the crucible to large temperature variations. As a result, both dissolution and devitrification are drastically reduced, thus increasing the life expectance of such crucibles by one order of magnitude. Further factors for improving the Czochralski technique are big charges, larger diameters or square-shaped rods[22, 23] respectively, and widening of specifications with respect to special requirements of solar cell application.

With these improvements in ingot production cost, ingot slicing becomes the focal point for further development. "Solar-grade" modification of the well-established ID sawing technique, based on lower requirements with regard to surface quality and thickness tolerance of the slices, is the first step. Enhanced ID slicing technology[24] is significantly increasing the number of usable slices per cm of ingot over industry practice. It requires reduction in both blade and slice thickness and can be accomplished by the combination of a new blade technology, cutting edge control and ingot rotation with a minimum of exposed blade area to reduce vibration effects.

3 Unconventional Refining of Silicon

Thermodynamic calculations[25] on the chemical preparation of elemental silicon from common silicon compounds indicate that nearly 150 out of 200 tested reactions are thermodynamically feasible. In about 20 cases feasible routes to solar-grade silicon could be demonstrated. This chapter deals with a relative small number of high-potential candidates in terms of low-cost processes.

For the production of the element, silicon dioxide and volatile silicon compounds represent the most widely used raw materials. Economically useful deposits of silicon dioxide – as the primary silicon source – can be found all over the world. Mining of quartzite rock and special strata of quartz sand requires little energy and has a negligible environmental impact.

Volatile silicon compounds represent the secondary silicon source. All volatile products have one property in common that can be of great significance, namely the ease in purification by fractionated distillation or desorption. Among the volatile halides silicon tetrachloride and tetrafluoride are the economically most useful compounds. Monosilane, as an intermediate product from the trichlorosilane conversion, can be considered a silicon source of tertiary character at most.

3.1 (Halo)Silane Conversion

In Fig. 5 the chemical potential, expressed by the temperature function of standard free enthalpy values ($-\Delta G^{\circ}_{298}$ – first-order approach), is plotted for different processes involving conversion of silicon halides into (high-purity) silicon. Any process reaching or exceeding the zero line from below is characterized by a high silicon yield. Whereas some reduction reactions of silicon tetrachloride can be run close to the zero line in the relevant temperature ranges, its reaction with sodium is highly exothermic. Figure 6 shows the

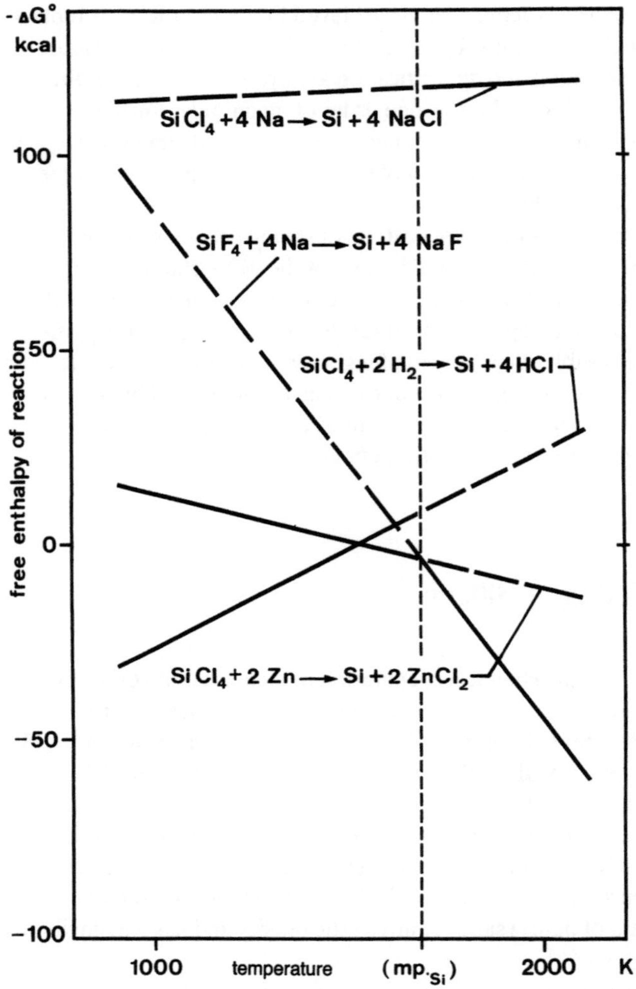

Fig. 5. Free enthalpy values of selected reactions for Si production from silicon tetrahalides[1)]

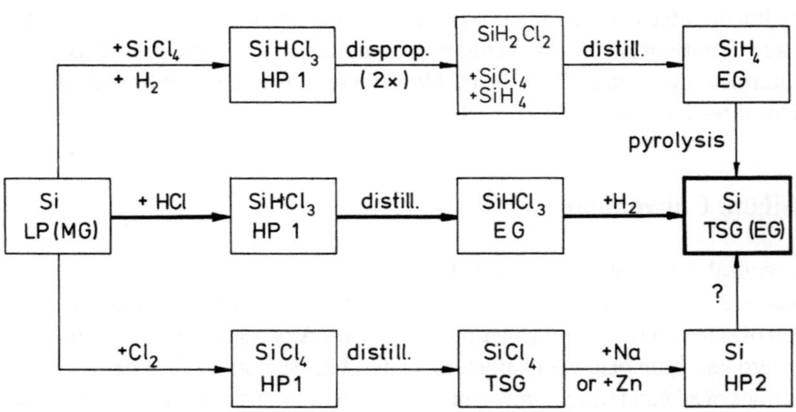

Fig. 6. "High purity" approaches to Si production via volatile chlorides (symbols see Fig. 4)[11)]

flow chart of different process sequences yielding high-purity silicon from or via silicon chlorides being generated from metallurgical-grade (MG) silicon. The symbols LP, HP 1, HP 2 and TSG indicate the purity categories of the different silicon products as quantified in Fig. 4. Starting from low-purity (MG) silicon, the symbols HP 1, HP 2 and TSG characterize basic, advanced and final stages of refining. In the center of Fig. 6 the conventional processing steps affording electronic-grade (EG) silicon are outlined: In the conventional Siemens "C-process" of today[27, 28] MG-Si is chlorinated by HCl to produce trichlorosilane, $SiHCl_3$. Impurities from MG-Si being carried over into trichlorosilane easily can be brought down to very low levels by fractionated distillation. During the subsequent vapor phase deposition on heated silicon rods, the hyperpure state of trichlorosilane is maintained in the polycrystalline material. The present expenses for silicon produced in existing semiconductor-grade polyplants are extremely high for terrestrial solar cells. Several manufacturers have proposed new processes that depart from conventional technology.

A low-cost process in which trichlorosilane is formed as an intermediate seems to be feasible. It involves conversion of silicon tetrachloride into monosilane and subsequent purification plus pyrolytic decomposition of the latter (see flow chart Fig. 7). This technology currently applied at Union Carbide[29, 30] utilizes SiH_4 generation by two disproportion reactions starting from trichlorosilane $SiHCl_3$ obtained from Si, $SiCl_4$ and H_2. The process carried out in a fluidized bed of 150 μm MG-Si powder proceeds as follows:

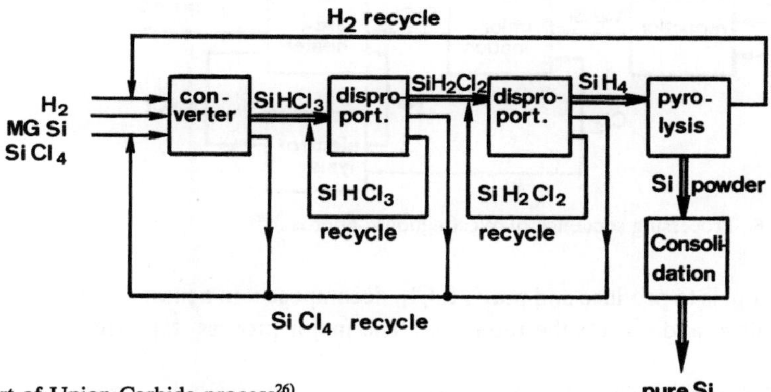

Fig. 7. Flow chart of Union Carbide process[26)]

$$Si(s) + 3\ SiCl_4(g) + 2\ H_2(g) \rightarrow\ SiHCl_3(g)$$

When using a copper catalyst, this reaction yields up to 30% molar conversion of $SiCl_4$ to $SiHCl_3$ on one pass at 500 °C and 30 atm. Liquid $SiHCl_3$ is then redistributed in a bed of a tertiary amine resin to give SiH_2Cl_2:

$$2\ SiHCl_3(l) \rightarrow\ SiCl_4(l) + SiH_2Cl_2(l)$$

This reaction performed at 55 °C and 3 atm yields a 10% molar conversion of $SiHCl_3$ to SiH_2Cl_2. SiH_2Cl_2 is disproportionated in a different catalyst bed as follows:

$$2\ SiH_2Cl_2\ (l) \rightarrow SiCl_4\ (l) + SiH_4\ (g)$$

This reaction has been demonstrated to yield about 12% molar conversion of SiH_2Cl_2 to SiH_4 on one pass at 60 °C and 3 atm. To produce silicon, the thermally unstable silane is decomposed at 700 °C in a free-space reactor:

$$SiH_4\ (g) \rightarrow Si\ (s) + 2\,H_2\ (g)$$

This pyrolysis yields a 90–100% conversion to Si powder. The resulting 0.1–1 μm large particles either have to be enlarged in a fluidized-bed reactor by using them as seed particles for further decomposition of SiH_4 to Si or to be consolidated directly via sintering or melting. To gain 1 kg Si, about 1000 kg $SiCl_4$ have to be recycled. Nevertheless, this technology is assumed to be one of the best investigated and, thus, a promising route to produce low-cost silicon on a large scale.

$SiCl_4$ generated by chlorination of MG-Si and purified by distillation has been examined as a source for high-purity silicon. The thermodynamically favorable reduction of $SiCl_4$ with sodium[31)], as depicted in Fig. 5, has recently been reinvestigated at Westinghouse[32, 33)]. As schematically shown in the flow chart in Fig. 8, the overall process

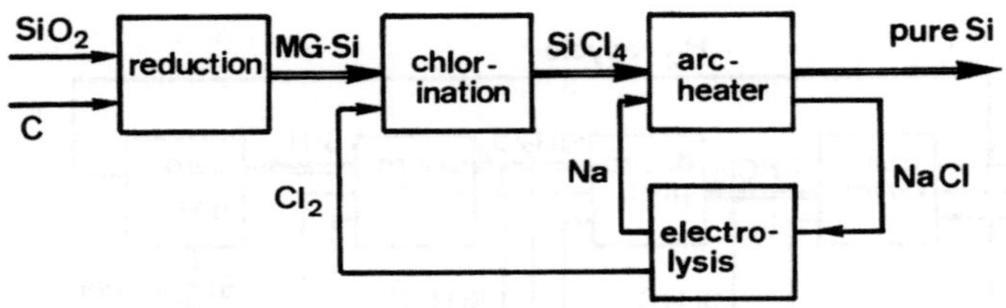

Fig. 8. Processing sequence of Westinghouse approach[26)]

concept is to produce and purify $SiCl_4$, decompose it in a plasma, collect molten Si droplets, and recycle the reductant. The major process steps are as follows:

$$MG\text{-}Si\ (s) + 2\ Cl_2\ (g) \rightarrow SiCl_4\ (g)$$

Vaporized Na and $SiCl_4$ are together injected into a plasma-heated mixture of hydrogen and argon at a temperature of about 2000 K and reacted according to

$$SiCl_4\ (g) + 4\ Na\ (g) \rightarrow Si\ (l) + 4\ NaCl\ (g)$$

The temperature has to be controlled so as to provide silicon as a liquid and the coproduct NaCl as a vapor. The mixture is passed through a cyclonic separator having a silicon-skull wall so that the silicon droplets are tangentially forced to the wall and then flow down to a molten mass. For a low-cost process the condensed NaCl is recycled by its electrolysis in Downs cells to provide chlorine for the production of $SiCl_4$ and sodium for its reduction.

An encouraging approach can be seen in the application of fluidized-bed technology to the so-called Du Pont process[34)] at Battelle[35)] and Rhone-Poulenc[36)]. The major steps are depicted in the flow chart in Fig. 9. The first step involves chlorination of MG-Si

$$MG\text{-}Si\ (s) + 2\ Cl_2\ (g) \rightarrow SiCl_4\ (g)$$

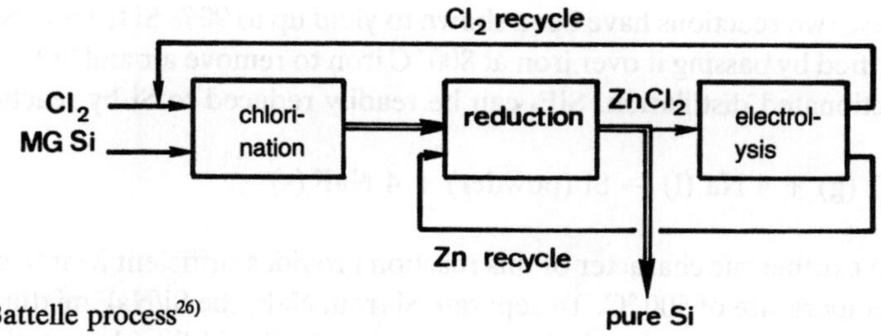

Fig. 9. Flow chart of Battelle process[26)]

This technical grade $SiCl_4$ is purified by distillation to attain solar-grade quality. Then, the reduction of $SiCl_4$ by zinc is initiated according to the reaction

$SiCl_4$ (g) + 2 Zn (g) → Si (particles) + 2 $ZnCl_2$ (g)

The gas-phase process takes place in a fluidized-bed reactor. The temperature must be higher than 918 °C (Zn condensation) but should not exceed 927 °C where the Si yield markedly begins to decrease due to $SiCl_2$ formation. The size of silicon seed particles increases from 500 μm to approximately 1000 μm. The Si yield at this rate is 62%, that is 90% of the thermodynamically calculated value. Unreacted $SiCl_4$ is recycled for purification and subsequent vaporization into the reactor for additional Si deposition. $ZnCl_2$ undergoes electrolysis to recycle Cl for the production of $SiCl_4$ and Zn for its reduction.

In Fig. 5 standard free enthalpy data for another reduction process are shown, the reduction of silicon tetrafluoride by sodium known as SRI process (Stanford Research Institute International)[37, 38)]. SiF_4 is formed as a by-product in the production of superphosphate fertilizer from phosphate ore. The ore contains CaF_2 which is converted to HF when treated with H_2SO_4. HF then reacts with SiO_2 to form SiF_4 with an estimated 75.000 t/a being recovered in the US as fluosilicic acid, H_2SiF_6. As shown by the flow chart in Fig. 10, NaF is added to H_2SiF_6 to form sodium fluosilicate:

H_2SiF_6 (aq.) + 2 NaF (s) → Na_2SiF_6 (s) + 2 HF (aq.)

After filtration and drying, Na_2SiF_6 is decomposed at 650 °C:

Na_2SiF_6 (s) → SiF_4 (g) + 2 NaF (s)

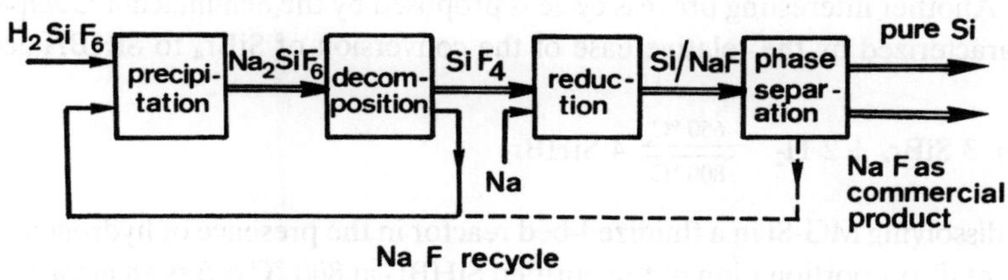

Fig. 10. Processing sequence of SRI approach[26)]

These two reactions have been shown to yield up to 90% Si from H_2SiF_6 and SiF_4. SiF_4 is purified by passing it over iron at 800 °C iron to remove air and SO_2 and by subsequent fractionated distillation. SiF_4 can be readily reduced to Si by reaction with sodium:

$$SiF_4\ (g) + 4\ Na\ (l) \rightarrow Si\ (powder) + 4\ NaF\ (s)$$

The exothermic character of this reaction provides sufficient heat to sustain the process at a temperature of 500 °C. To separate Si from NaF, the Si/NaF mixture subsequently is melted forming two immiscible phases whereby liquid-liquid extraction provides additional purification. As the reduction reaction yields four times as much NaF as Si, low-cost Si is attainable by adequate by-product credit only.

Another process based on [Si-F] has been investigated at Motorola[39, 40]. Its first step involves the reaction of silicon tetrafluoride with MG-Si at high temperatures (1100 °C) to yield gaseous silicon difluoride in a fluidized-bed reactor:

$$MG\text{-}Si\ (s) + SiF_4\ (g) \rightarrow 2\ SiF_2\ (g)$$

This step is followed by association of gaseous SiF_2 at low temperature (− 80 °C):

$$x\ SiF_2\ (g) \rightarrow (SiF_2)_x,\ x = 2 - 11$$

In the course of this reaction the majority of the impurities that were present in the original MG-Si and which were transported by the SiF_2 gas is reported to be removed. In the final step the polymer is decomposed at about 250 °C and 150 torr:

$$(SiF_2)_x \rightarrow \frac{x}{2}\ Si\ (powder) + \frac{x}{2}\ SiF_4\ (g)$$

Further heating removes residual fluorides from the Si powder. SiF_4 is recovered to react again with MG-Si.

The objective of a non-equilibrium plasma process at Aerochem[41] is to generate hydrogen atoms for the reduction of $SiCl_4$.

$$SiCl_4\ (g) \xrightarrow[-\ HCl]{+\ H} SiH_4\ (g)\ \text{or}\ Si\ (s)$$

Experiments have shown the possibility to produce up to 3% atomic hydrogen. $SiCl_4$ injected into this plasma-like system reacts at 400–600 °C to yield Si powder. As far as monosilane is generated, it can be seen as an additional silicon source (see e.g. UCC process).

Another interesting process cycle is proposed by the Schumacher Company[42]. It is characterized by the relative ease of the conversion of $SiBr_4$ to $SiHBr_3$ according to

$$Si + 3\ SiBr_4 + 2\ H_2 \underset{800\ °C}{\overset{650\ °C}{\rightleftharpoons}} 4\ SiHBr_3$$

by dissolving MG-Si in a fluidized-bed reactor in the presence of hydrogen. The subsequent disproportionation of the purified $SiHBr_3$ at 800 °C offers an attractive potential in terms of a cyclic process.

3.2 SiO_2 Conversion

Silicon dioxide is composed of the two most abundant elements in the earth crust. In nearly all geological periods SiO_2 has been concentrated in vast quartz deposits all over the world. The different SiO_2 qualities are compiled in Fig. 11. The quality increases from impurity concentrations of about 1% in quartzite rock to <0.03% in quartz veins or rock crystals. Medium-purity SiO_2 is found in chemically converted special silicates and in quartz sands, either genuine or purified by chemo-mechanical treatment. Prices of quartz products reach from 0.02 $/kg for quartz sands up to some 10 $/kg for rock crystals. The preference or necessity to use a certain quality depends largely on the type of reduction and sequence of refining processes.

mining product	quartzite rock	special silicates	quartz sand	purif.sands pegmatites	quartz veins rock crystal
up-grading	standard (tech.grade)	chemical conversion	standard (tech.grade)	chem.mech. treatment	standard
nature of particles	chunks	powder	grains	grains	nuggets
impurities % wt.	~ 1	< 0,3	>0,2	< 0,05	< 0,03

SiO_2 Qualities

Fig. 11. Different types of SiO_2 sources[1)]

Besides the well-known carbothermic reduction, a more economical route involves direct preparation of silicon from its dioxide, i.e. an aluminothermal process leading to silicon-aluminium alloys. In Fig. 12 the chemical potential, expressed by the temperature function of the standard free enthalpy value, is depicted for these two reduction reactions.

$$SiO_2 + 2\,C \rightarrow Si + 2\,CO$$

$$3\,SiO_2 + 4\,Al \rightarrow 3\,Si + 2\,Al_2O_3$$

Whereas the carbothermic reduction of SiO_2 is inattractive below 2000 K, the reduction with Al is a pronounced exothermic reaction over the whole temperature range.

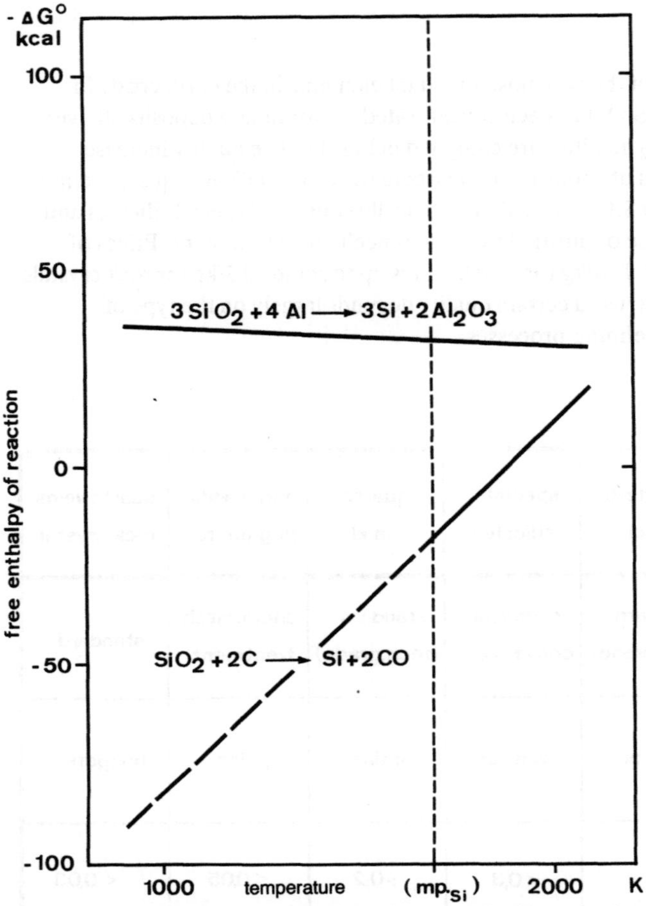

Fig. 12. Free enthalpy values of selected systems for Si production from SiO_2[1)]

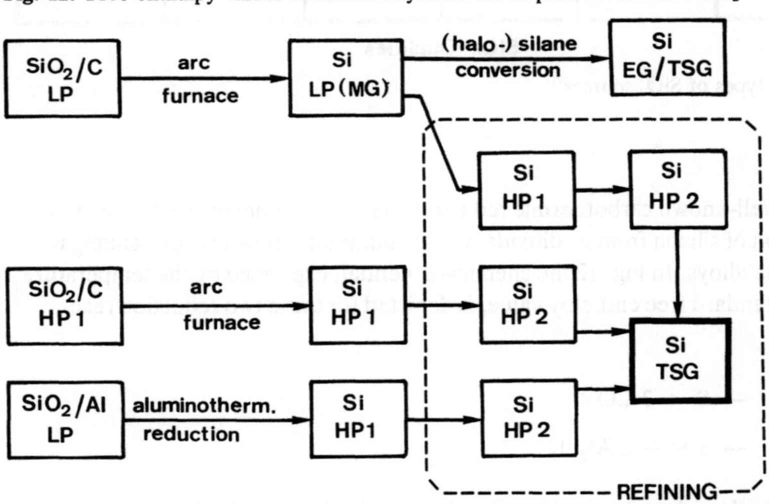

Fig. 13. Different approaches to large-scale production of Si for terrestrial solar cells (symbols see Fig. 4)

In Fig. 13 three different routes to silicon are schematically described. The two processes using carbon as reductant differ mainly in the quality of the starting materials. As the individual steps designated symbols LP, HP 1 and HP 2 indicate, the LP process requires more severe purification to reach the TSG quality level. The different refining concepts are treated in Sect. 3.3.

The only large-scale process which is technically well established since the turn of the century is the reduction of silicon dioxide with carbon in an electric arc furnace (Fig. 14). This low-purity process yields metallurgical-grade silicon (98% Si) with an estimated worldwide capacity of 500,000 tons per year. Average analyses of metallurgical-grade silicon samples from different suppliers show that it contains about 0.5% Fe, 0.4% Al, 100–400 ppm each of Ca, Cr, Mg, Mn, Ni, Ti, and V, and 20–40 ppm each of B, Cu, P, and Zr[43].

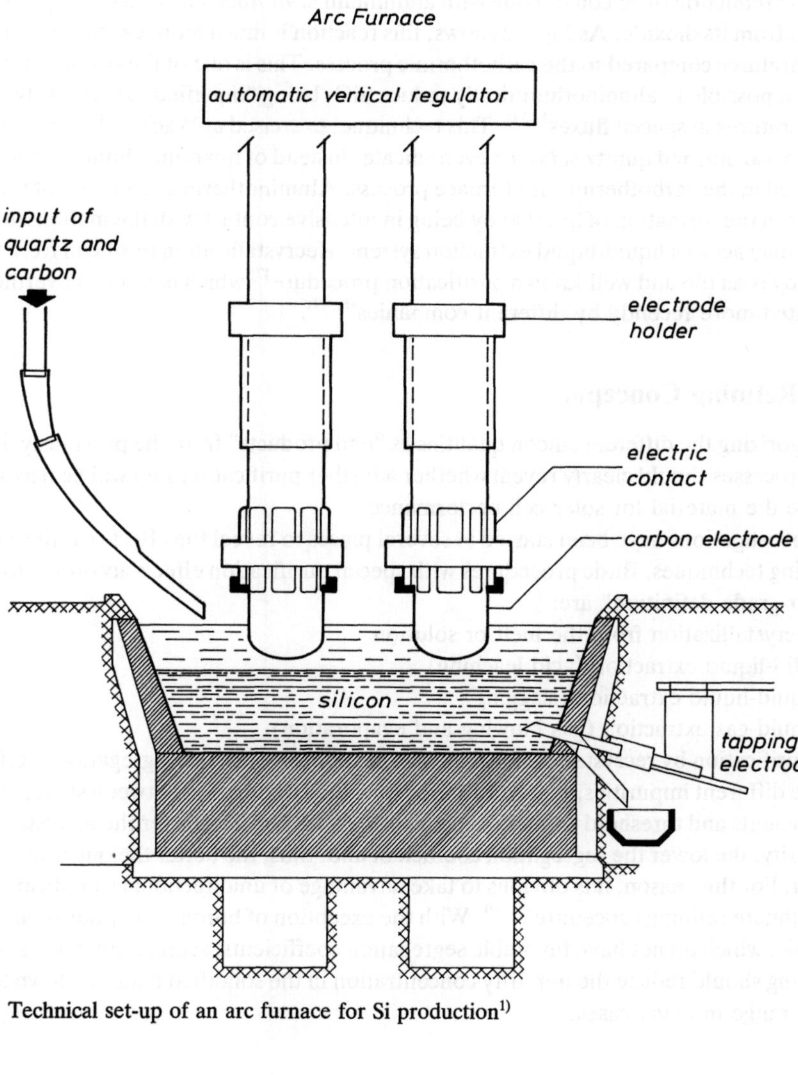

Fig. 14. Technical set-up of an arc furnace for Si production[1]

Quite recently, efforts have been made at Dow Corning[44–46] to drastically lower the impurity level of arc-furnace silicon. The improved and expanded metallurgical-grade silicon technology demands the use of high-purity raw materials in a specially designed arc furnace, the Direct Arc Reactor (DAR).

Vast quartz deposits have been identified meanwhile as adequate sources of silica with respect to purity. As reductant, carbon black powder formed into pellets (sucrose as binder) is used. Improved arc-furnace materials, construction and operation are utilized to maintain purity requirements. The purity of typical DAR-silicon is 99.99%. Residual impurities are Al and Fe in the range of 50–100 ppm each, B, P and other metals below 10 ppm each. To reach TSG quality, final purification by extractive refining and/or directional solidification is required. This material has been claimed to offer the potential of being used either as a substrate for epitaxially finished solar cells, to provide silicon for polycrystalline cells using casting (and sawing) techniques, or to produce silicon for directly grown ribbons.

The reduction of silicon dioxide with aluminium is another way of directly preparing silicon from its dioxide. As Fig. 12 shows, this reaction is much more exothermic at lower temperatures compared to the carbothermic process. This is one of the conditions that make it possible to aluminothermically reduce SiO_2 being "liquefied" at relatively low temperatures in special fluxes[1, 11]. This technique, exercised at Wacker, has the advantage to use purified quartz sands or even silicates instead of quartzite chunks as they are required in the carbothermic arc-furnace process. Aluminothermic reduction of this kind results in the formation of Si-Al alloy being in intensive contact with the molten flux. The latter may act as a liquid-liquid extraction system. Recrystallization of silicon from the Si-Al alloy is an old and well-known purification procedure[47] which has been nevertheless patented more recently by different companies[48, 49].

3.3 Refining Concepts

Categorizing the different silicon qualities as "end products" from the previously discussed processes should clearly reveal whether a further purification step will be necessary to utilize the material for solar cell performance.

Investigations have been started at several places to reveal the effect of different refining techniques. Basic procedures with special purification effects according to "solar-grade definition" are:

- Recrystallization from the melt or solution
- Solid-liquid extraction (acid leaching)
- Liquid-liquid extraction (slagging)
- Liquid-gas extraction (gas blowing and evaporation).

Purification by recrystallization from the melt depends on the segregation coefficients of the different impurities. Figure 15 shows a striking correlation between segregation coefficients and threshold value for solar cell degradation: The lower the admissable impurity, the lower the segregation coefficient and, thus, the better the purification effect. For this reason, it is obvious to take advantage of unidirectional solidification as an ultimate refining procedure[48, 50]. With the exception of boron, phosphorus, and arsenic, which do not have favorable segregation coefficients, segregation during normal freezing should reduce the impurity concentration in the solidified material down to the ppba range in many cases.

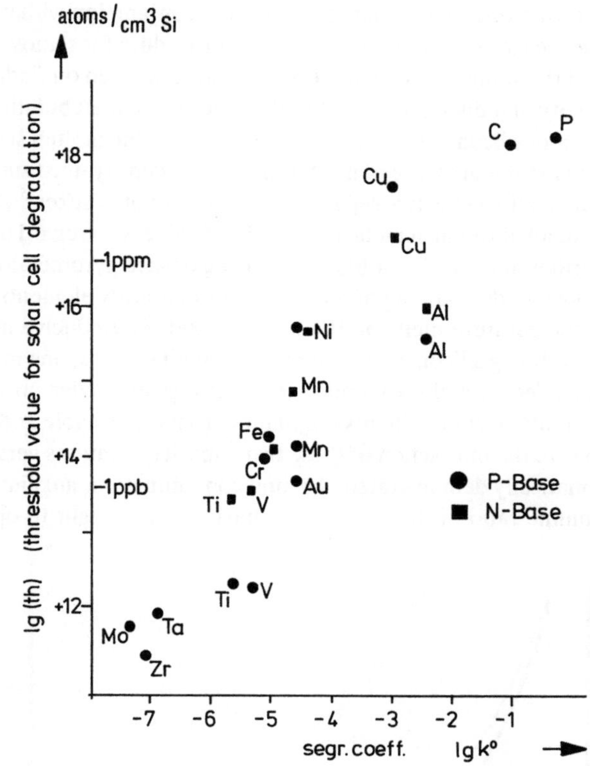

Fig. 15. Correlation of solar cell degradation and segregation coefficient for different impurities in Si[10)]

Impurities like iron and aluminum which are present in MG-Si at higher levels concentrate at the grain boundaries due to excessive impurity enrichment in the residual melt which exceeds the effective solubility in solid silicon[51)]. Impurity inclusions at the grain boundaries of polycrystalline MG-Si in the form of silicides or metal eutectics are a requisite to hydrometallurgical upgrading. Purification of MG-Si by acid leaching has been reported over the last half century[52–55)]. Since impurities with favorable segregation coefficients concentrate at the grain boundaries of the polycrystalline material, MG-Si lumps have to be crushed and milled to powder. Leaching with solutions of hydrochloric[56)], hydrofluoric, sulfuric[57)] and nitric acid, respectively, easily reduces metal impurities by more than one order of magnitude[58)]. As indicated by multielement analysis[59)], impurities like Fe, Al and Ca can be eliminated by solid-liquid extraction down to low ppm concentrations whereas in the case of Mn, Ti, Cr, and Ni even sub-ppm levels have been observed.

Blowing a reactive gas through molten silicon has been described as another effective procedure for removing impurities. The level of Al and Ca, for example, can be reduced by blowing with chlorine or oxygen, respectively[60)]. Blowing with CO_2 is reported to additionally lower the carbon content[61)]. To reduce the phosphorus concentration, MG-Si melts have been successfully treated with O_2, H_2O, CO and CO_2[62)]. The gases listed above were used in single to three-component mixtures together with H_2, SiF_4 and SF_6 in another experimental set-up[63)].

The use of slagging – either alone or in conjunction with gas blowing – has been suggested as another pyrometallurgical procedure for removing impurities[64–67]. The success of this liquid-liquid extraction technique is based on finding a slag that has a higher solubility of individual impurities than the silicon melt but must not dissolve a significant amount of silicon itself. It is also necessary that the molten slag exhibits high chemical stability in the presence of molten silicon. Its density must sufficiently differ from that of silicon to allow effective separation. As proven at Wacker[55], liquid-liquid extraction with slags of alkaline earth metal silicates is effective with regard to strongly electropositive impurities like Al, Ca, or Mg. Optimizing of such pyrometallurgical processes requires a thorough understanding of the behavior of impurity elements in liquid-liquid connected high-temperature melts on the basis of oxidative/reducive interactions.

In refining silicon by treating it in the molten state, one must consider that besides the gradual decrease of the impurity level through extractive purification one has to deal with an increase of contaminants originating from the crucible and other auxiliary materials. In Fig. 16 the interaction of these two impurity situations versus time of extraction is schematically demonstrated. To limit contamination, auxiliary materials and the surface to volume ratio of the silicon melt have to be brought to optimum conditions.

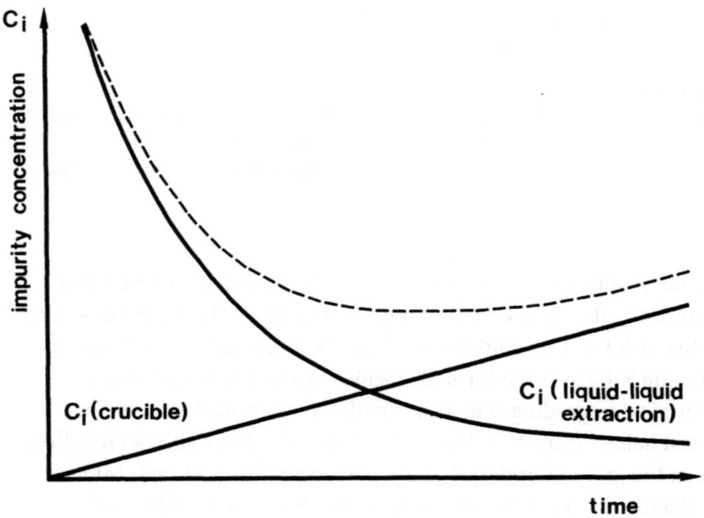

Fig. 16. Extractive purification of Si in the molten state and contamination from auxiliary materials

The final combination of processes required to upgrade silicon will not necessarily involve all the steps discussed. Figure 17 indicates feasible refining procedures leading from different starting qualities of solid material via HP 1 silicon and HP 2 to the final stage of TSG quality. Each of the steps in the refining sequence is designed to remove specific impurities and thus improve the electrical quality of the material.

MG-Si requires one first upgrading step[68] by solid-liquid extraction before achieving the HP 1 category of Si generated by carbothermic processing based upon high-purity starting materials (DAR-process, Dow Corning). Alternatively to acid leaching, MG-Si can be purified in a first step by recrystallization from Al-solution[49, 69]. This Al-based

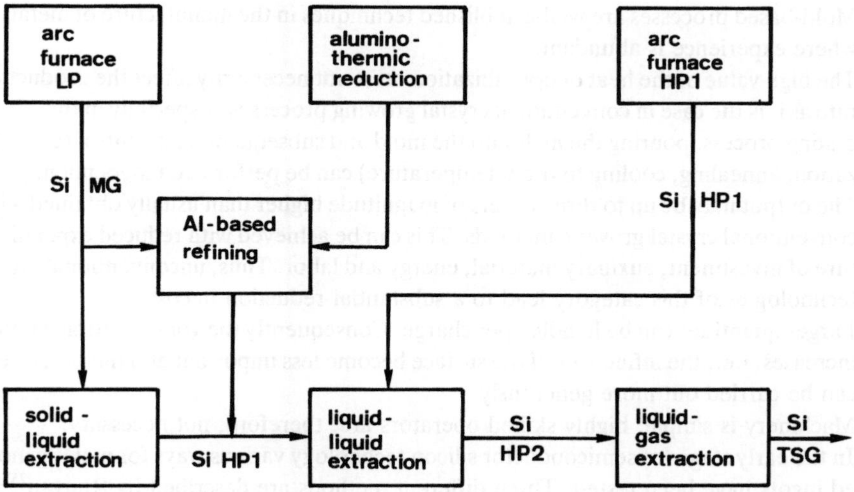

Fig. 17. Different concepts for refining Si to terrestrial solar-grade quality (symbols see Fig. 4)

refining is also under investigation at Wacker[9] as an integrated step in the sequence of an aluminothermic reduction concept. To reach HP 2, a further refining procedure is needed (e.g. liquid-liquid extraction) as indicated in Fig. 17. According to the type of solar cell and device processing, a final purification step by liquid-gas extraction, or unidirectional solidifiaction may be applied[68, 70].

The process sequences discussed above represent only a small number of the broad variety of possible refining concepts. Differences in refining concepts may be caused by different auxiliary materials, by more or less improved starting silicon qualities or, last not least, by accidentally existing technical equipments. For example, one of these different refining concepts[71] starts with liquid-liquid extraction of MG-Si with a barium oxide-based slag in conjunction with oxidative gas blowing and ends up with a hydrometallurgical step. Another purification procedure[72–74] involves the following sequences: Precipitation of Si platelets from a solution of MG-Si in molten Al, acid leaching of the Si platelets, melting of the platelets in contact with silica slag, and directional solidification of the silicon-slag melt.

To design a basic low-cost refining concept, cost, availability, and purity of the materials to be used, utilization or recycling of by-products, and the amount of investment as well as energy consumption have to be considered. The simpler the basic process and the lower the number of necessary processing steps, the better are the chances to reach this target.

4 Unconventional Crystallization Techniques from the Melt

4.1 Ingot Technology

The conditions for silicon base material in terms of geometry, crystal perfection, production costs, and material throughput can be realized principally by unconventional ingot technologies as outlined below:

a) Mold-based processes are well-established techniques in the manufacture of metals where experience is abundant.
b) The high value of the heat of crystallization must not necessarily affect the production rate as it is the case in conventional crystal growing processes. Especially in the casting process, pouring the melt into the mold and subsequent heat control (crystallization, annealing, cooling to room temperature) can be performed independently. The output may be up to three orders of magnitude higher than usually obtained with conventional crystal growing methods. This can be achieved with reduced expenditure of investment, auxiliary material, energy and labor. Thus, unconventional ingot technologies of this category lead to a substantial reduction in cost.
c) Larger quantities can be handled per charge. Consequently the volume/surface ratio increases, i.e., the influences of the surface become less important and the processes can be carried out more generously.
d) Machinery is simple; highly skilled operators are, therefore, not necessary.

In the early stages of semiconductor silicon technology various ways for making mold-based ingots have been tested. Three different methods are described by Runyan[75]:

1. Ambit casting – in this process a cold plug is immersed in a silicon melt where a silicon disk immediately starts to be formed according to heat extraction conditions. This process shows some characteristics of liquid phase epitaxy and is applied in a modified form to prepare silicon layers on substrates (see Sect. 4.2.2)
2. Directional solidification or Bridgman-Stockbarger crystal growth is an effective and simple technique used for manufacturing of metal and alkali halide crystals[76]. Directional solidification has become an economically attractive approach to large-scale production of silicon ingots for terrestrial solar cell application.
3. Another method consists of pouring silicon into a preheated or cooled mold and subsequent (directional) solidifiaction. This is a casting technique in its original sense (filling a melt into a form or die) and can be practised either batchwise or continuously.

The most common variations of directional solidification and casting techniques are listed in Fig. 18. The different methods are characterized in particular in the subsequent paragraphs.

UNCONVENTIONAL
Ingot Technology

development in silicon"casting" techniques

die material	crystallinity of ingot	company
quartz,one-way mould	mono	Crystal Systems /77/
ceramic,one-way mould	poly	Union Carbide /50/
? / one-way mould	poly	Solarex /78,79/
graphite,one-way mould	poly	IBM /80/
graphite,reusable mould	poly	CGE /81/
graphite,reusable mould	poly	Wacker /82,83/
graphite die,continuous casting	poly	Wacker /84/

Fig. 18. Unconventional ingot technology: Development in casting and related techniques

4.1.1 *Directional Solidification*

The main problem in this field is the high reactivity of molten silicon by which almost every material is wetted. Contamination of a silicon melt by container materials can be quite severe. In most cases, the solidifying melt adheres to the container wall and, as a consequence of different thermal contraction, usually causes both silicon ingot and container to shatter.

Requirements for Container Materials

Materials used in contact with molten silicon must exhibit a high degree of inertness against the latter. Due to a considerable residence time of the melt, dissolution of the container material hardly can be completely suppressed. As container materials can be used high-density graphite[80, 81)], quartz "ceramics"[77)] and mullite[85)]. Both quartz and graphite parts are today available in high-purity grades, a fact which may cause minor problems only in terms of additional contamination (besides O and C). Kazmerski et al.[86)] have shown by SIMS studies that the corresponding spectra taken from samples of different silicon ingots were characteristic of the container materials used.

In the absence of particular precautions the solidifying melt adheres to the container walls. During the cooling process, in such a case, either the ingot or the container (or both) will crack due to differences in the thermal expansion coefficients. T. F. Ciszek[80)] has used dense types of graphite with adjusted physical properties (e.g. coefficient of thermal expansion = CTE). In the case of mullite, $(SiO_2)_x(Al_2O_3)_{1-x}$, too, CTE adjustment has been reported[85)]. Schmid[79)], on the other hand, has utilized mechanically instable quartz crucibles by choice. Their walls are thin enough to fracture during solidification of silicon before the ingot starts to crack.

The methods mentioned generally use one-way molds. Similar to[80)], Fally et al.[81)] have used graphite containers which, by proper handling, served their purpose several times. Another way to take care of the cracking problem is the protection of the inside of the container by powders of Si_3N_4[82)], SiO_2, Si, and graphite (see also[75)]).

Solidification

Generally, one can differentiate between two basic techniques:

a) The container is moved within a fixed temperature profile causing growth of solidified mass from bottom to top (see Fig. 19). This principle has been utilized by Ciszek et al.[80)], Kotval et al.[50)], and Fally et al.[81)]. Crystallization may be initiated either by seeding or spontaneous nucleation. So far, monocrystalline growth could not be maintained throughout the total solidification process[80)].

b) Both container and heating system are in fixed position and the temperature profile is varied by reducing the heating energy[80)] (see Fig. 20). An interesting modification of this principle has been published by Schmid[77)]. In his Heat Exchanger Method (HEM) the gradual reduction of the heating current is combined with a simultaneously localized heat extraction by a helium jet stream (see Fig. 21). The latest version of this kind[88)] contains a seed (mono) crystal in direct contact with the gas-cooled heat exchanger, thus facilitating the control of the different stages of monocrystalline growth.

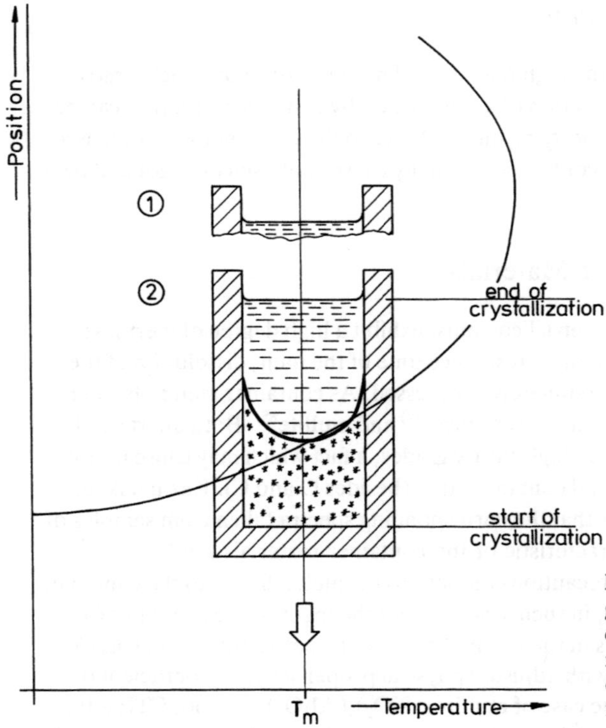

Fig. 19. Schematic graph of ingot growth by directional solidification or Bridgman-Stockbarger method. The crucible is moved downward in a fixed temperature field

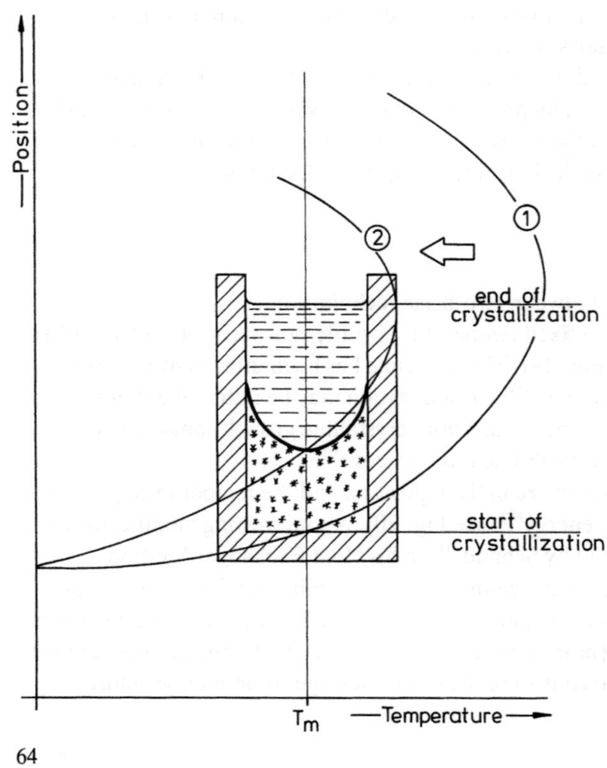

Fig. 20. Schematic graph of ingot growth by directional solidification. The crucible is held fixed while the temperature profile is "moved" through the crucible

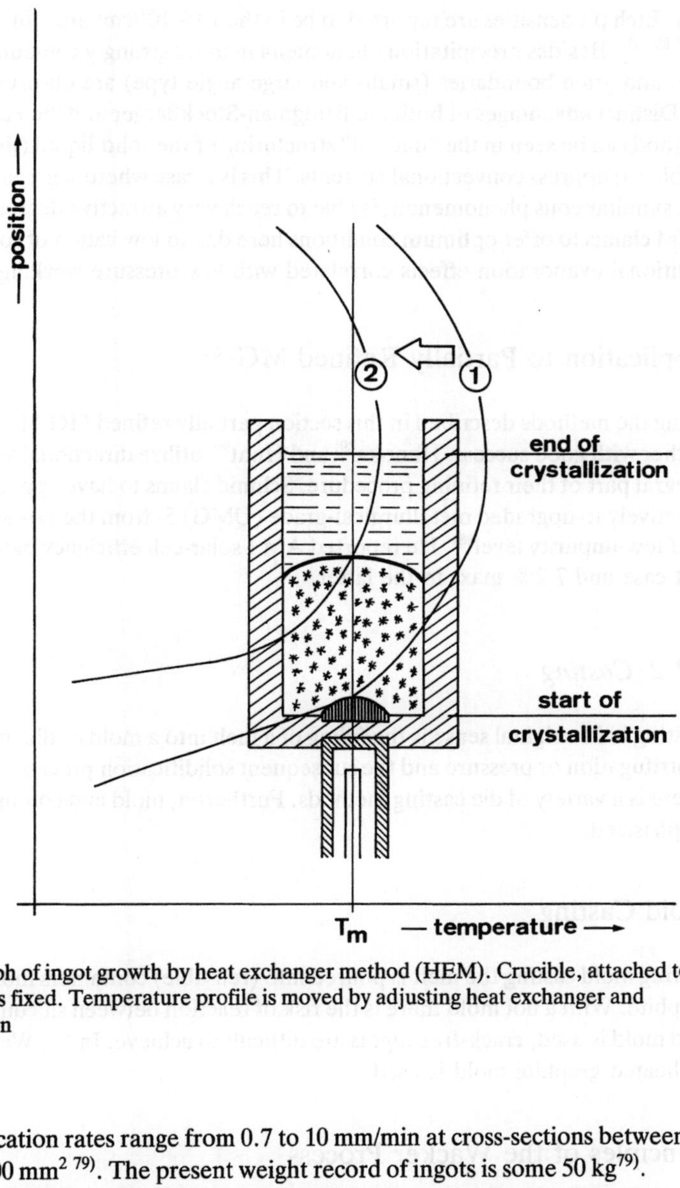

Fig. 21. Schematic graph of ingot growth by heat exchanger method (HEM). Crucible, attached to heat exchanger, remains fixed. Temperature profile is moved by adjusting heat exchanger and heating power operation

Reported solidification rates range from 0.7 to 10 mm/min at cross-sections between 50 × 50 and 300 × 300 mm^2 [79]. The present weight record of ingots is some 50 kg[79]. Crystallization is kept intentionally at low rates (0.5–6.0 kg/h, depending on the cross-section) and is at present "standardized" toward 20 kg, representing the final ingot.

Properties of Silicon Ingots

Besides the speciality of the largely monocrystalline ingots grown by HEM, a typical feature of the directionally solidified silicon is a rather coarse grain structure. Depending on the crystallization conditions, a more or less pronounced columnar pattern along the main vector of solidification is observed. Typical crystallite diameters range from 1 to 20 mm. Etch pit densities are reported to be in the 10^3–$10^4/cm^2$ area, in some cases up to $10^6/cm^2$ [80, 81]. Besides precipitation phenomena in more strongly contaminated material, twin and grain boundaries (small- and large-angle type) are observed.

Distinct advantages of both the Bridgman-Stockbarger and the Heat Exchange Method can be seen in the "natural" structuring of the solid-liquid (thermal) system that is able to suppress convectional currents. This is a case where segregational purification, as a simultaneous phenomenon, is able to reach very attractive degrees of efficiency. HEM claims to offer optimum conditions here due to low values of solidification rate and additional evaporation effects correlated with low-pressure working conditions.

Application to Partially Refined MG-Si

Using the methods described in this section partially refined MG-Si has been purified further with good success. Hanoka[89] and Hunt[90] utilize directional solidification as an integral part of their refining procedure. Schmid claims to have applied his method quite effectively to upgraded metallurgical-grade (UMG) Si from the DAR reactor[91] and MG-Si of low-impurity level[88]. He reported AM 1 solar-cell efficiency data up to 12.3% in the first case and 7.2% max. in the latter.

4.1.2 Casting

Casting in its original sense is the filling of a melt into a mold or die by the aid of gravity, centrifugation or pressure and the subsequent solidification process within the container. There is a variety of die casting methods. Furtheron, mold and continuous casting will be emphasized.

Mold Casting

During mold casting the melt is poured into (reusable) containers mostly consisting of graphite. With a hot mold there is the risk of reaction between silicon melt and mold. If a cold mold is used, crack-free ingots are difficult to achieve. In the Wacker process[82, 83] a preheated graphite mold is used.

Principles of the Wacker Process

Figure 22 shows the main part of the set-up which is used for casting multicrystalline silicon ingots. The starting material (high-purity semiconductor-grade silicon or refined metallurgical-grade silicon[93]) is melted in a quartz crucible (1) with a graphite support (2) around it. The silicon melt is poured through a graphite supported (5) quartz funnel (4) and fed into a graphite mold (8) that was preheated appropriately. Rotating the mold ensures an even distribution of the heat, which helps to develop a homogeneous chill layer of solid silicon before the silicon melt is able to react with the mold walls. The silicon melt can also be poured directly into the mold without using a funnel, but a thermally stabilized funnel helps to balance inhomogeneities in the silicon melt. The

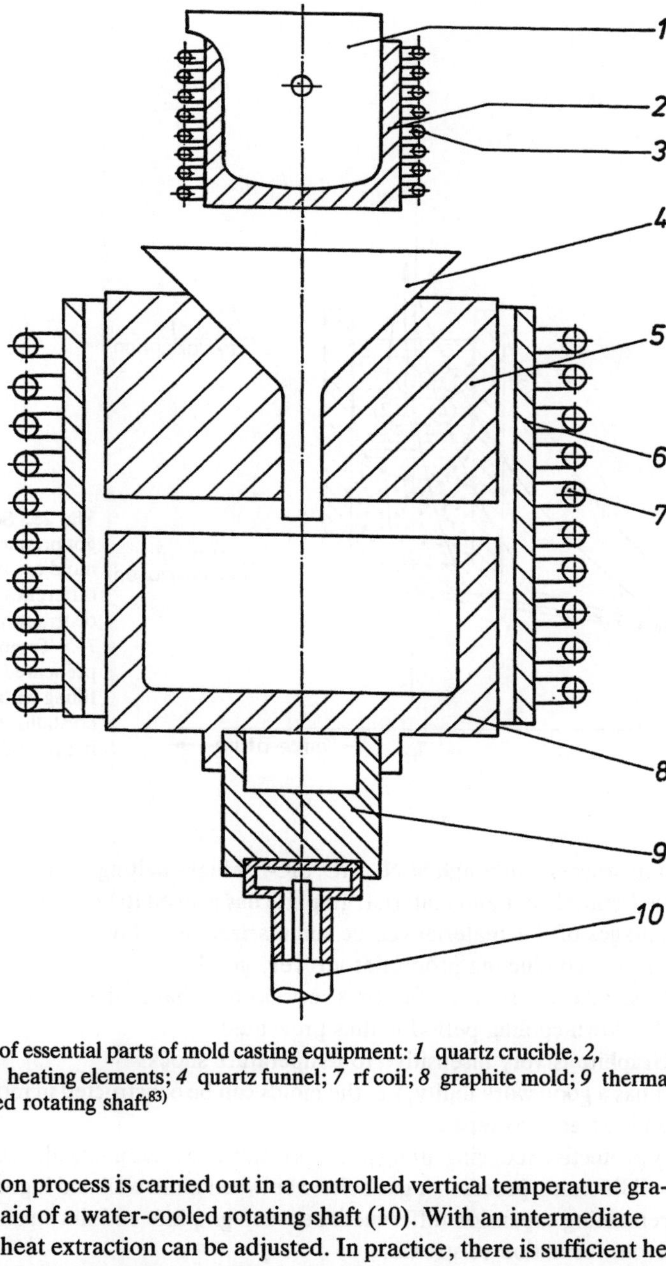

Fig. 22. Schematic view of essential parts of mold casting equipment: *1* quartz crucible, *2, 5* graphite supports; *3, 6* heating elements; *4* quartz funnel; *7* rf coil; *8* graphite mold; *9* thermal isolation; *10* water-cooled rotating shaft[83])

subsequent solidification process is carried out in a controlled vertical temperature gradient, adjusted by the aid of a water-cooled rotating shaft (10). With an intermediate thermal insulation (9) heat extraction can be adjusted. In practice, there is sufficient heat loss from the free surface of the melt for a solid crust to form. In this case, the last region to solidify is in the interior. As silicon expands during freezing, melt inclusion leads to a highly stressed zone or to cracking of the ingot. It is, therefore, necessary to keep the melt surface open until the solidification process is finished. Temperature profiles inside the mold are demonstrated in Fig. 23 for the different processing stages.

Most important for the quality of the ingot is the mold that must be adjusted in terms of material and construction to the nature of the molten material and to the individual

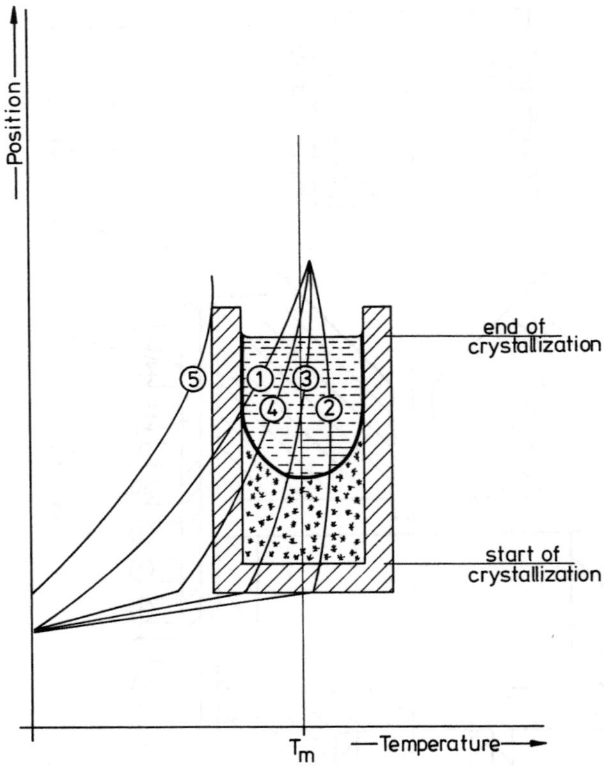

Fig. 23. Schematic graph of ingot preparation by reusable-mold casting technique. The different stages of temperature distribution are given by curves *1–5*. Temperature profiles of *1* preheated empty mold; *2* after filling with silicon melt; *3* during crystallization; *4, 5* during cooling period

casting process. Although Si_3N_4, SiC, SiO_2 or high-melting ceramics like mullite have proved suitable as mold material, graphite has proved to be the most qualified one. The advantages of this material can be summarized as follows:

1. Its heat-conducting properties are very good
2. Its small expansion coefficient is very close to that of silicon; cracking of ingots during the down-cooling period is thus prevented
3. Graphite is very insensitive to temperature shocks
4. It has a good workability, i.e. the molds can be constructed in more complex manner and are easy to repair
5. A protective reducing atmosphere is established automatically during the casting and solidification process. A gas bolster between mold wall and ingot helps to balance the remaining expansion differences between graphite and silicon and, as a consequence, avoid formation of stresses

In the meantime, a casting process of this kind is carried out on a larger scale in a pilot line[94]. A schematic view of the set-up for the semicontinuous production of silicon ingots is shown in Fig. 24. Additional components are: a recharging chamber, a cooling chamber as well as a loading/unloading chamber for withdrawing the mold with solidified ingot and introducing an empty mold. The cast ingots preferably have a square shaped cross-section of 106 × 106 mm^2 or 162 × 162 mm^2 and a weight of up to 8 kg. An enlargement of these values will be realized soon.

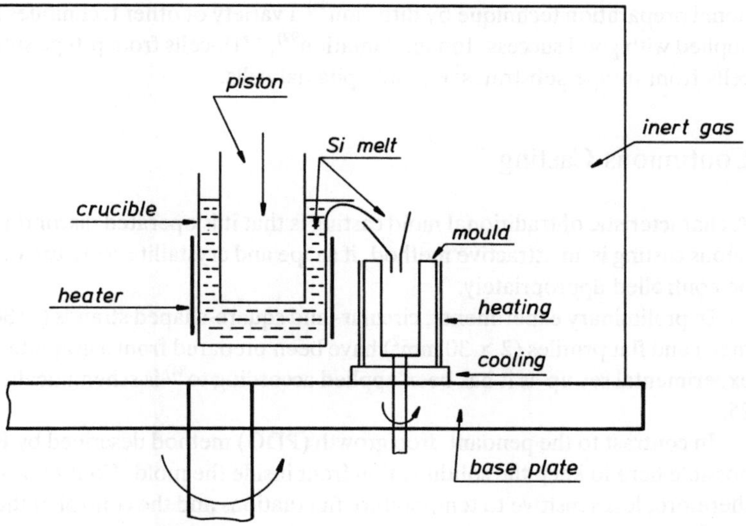

Fig. 24. Schematic view of an equipment for semicontinuous production of multigrain silicon ingots by mold casting[9)]

Nucleation in Mold Casting

The crystallites that were nucleated initially at the bottom and on the walls of the mold are usually randomly oriented. They grow in competition with each other and a preferred orientation develops as a result of the suppression of those crystals that are least favorably oriented toward the heat flow vector[95)]. A rapid lateral grain growth seems to be at least as important as the rapid growth perpendicular to the mold wall (e.g. to avoid enclosed melt droplets). The elimination of unfavorably oriented grains is gradual; increase of average grain size and "sharpening" of the orientation develop simultaneously. If the same steep thermal gradient persists throughout the solidification process, a columnar structure proceeds to grow.

In an experimental set-up, the temperature curves of a solidifying silicon melt in a graphite mold (162×162 mm^2 cross section) have been measured. The temperature curves show that the ingot solidifies at a rate of 85 g/min. It seems not worthwhile to increase the throughput by increasing the growth rate, because high growth rates result in very fine grain structures. In large-scale production, growth and cooling velocities must not affect the production rate: crystallization and cooling can be handled independent of the melting and casting stages.

Photovoltaic Performance

Experiences in recent years have shown that multigrain silicon (SILSO) is qualified as a base material for solar cells with very promising characteristics and quite acceptable conversion efficiencies of about 10% AM 1 at least[96, 97)]. The material is not very sensitive to the process used for the preparation of terrestrial solar cells. Besides the conventional preparation technique by diffusion[98] a variety of other techniques have been applied with good success: Ion implantation[99], MIS-cells from p-type substrates[100], ITO-cells from n-type substrates[101], and epitaxial cells.

Continuous Casting

A characteristic of traditional mold casting is that it is operated discontinuously. Continuous casting is an attractive method, if shape and crystallite structure of the strands can be controlled appropriately.

In preliminary experiments, circular- and square-shaped strands (∅ 30 mm, 30×30 mm^2) and flat profiles (2×30 mm^2) have been prepared from a graphite mold. An experimental set-up as it has been applied according to[84] is schematically depicted in Fig. 25.

In contrast to the pendant drop growth (PDG) method described by Ricard[102], it is possible here to keep the solidification front inside the mold. Continuous casting is, therefore, less sensitive to temperature fluctuations and the control of the solidification

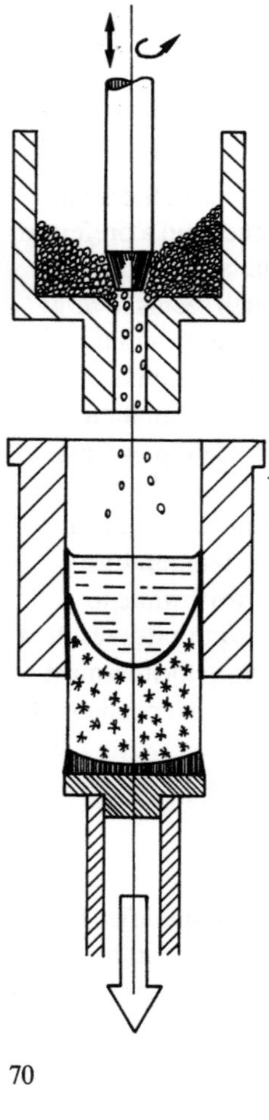

Fig. 25. Schematic graph of ingot preparation by continuous casting. Silicon is fed from a reservoir to the crystallization chamber. Solidification rate of ingot is adjusted to temperature profile to keep solid/liquid interface at same level

front must not be so stringent. The feasibility of continuous melt replenishment and the possibility of seeding (seed with columnar structure) have been demonstrated. Grain boundaries are predominantly arranged in the growth direction. The temperature gradients perpendicular to the growth direction are rather shallow and the crystallization front is thus quite planar. An increase in growth velocity (up to 10 mm/min had been achieved) necessitates steeper temperature gradients which result in a more disturbed crystal structure.

The relatively poor performances of solar cells fabricated from these materials showed that there are still problems with the grain structure. Individual grains have a size of several hundred microns at least, but sometimes a crystal pattern similar to the "grit structure" appears in the axial part of a strand. This grit structure has been found to impair the diode characteristics drastically. Although early results look very promising, further work has to demonstrate the potentialities of continuous casting as a method for low-cost mass production and the applicability of this material in terrestrial photovoltaics.

4.1.3 Unconventional Slicing

To be cost-effective, any ingot technology has to be coupled with an efficient slicing method. Similar to material refining and bulk crystallization, costs for slicing have to be reduced by a factor of 20–30, too. Material waste is another factor by which silicon sheet costs are influenced. However, very low values for kerf loss and slice thickness are no prerequisites here if add-on cost for material refining and bulk crystallization can be kept sufficiently low.

A recent comparison[103] of various slicing techniques has revealed that ID sawing shows the highest probability to reach the cost goal of the US Photovoltaic Program, due to its technological maturity (see also[104]). "Solar-grade" optimization of this well-established sawing technique is being attempted[24] (see also Chap.2).

Nevertheless, other forms of cost prediction show that ID sawing, in more advanced stages of PV technology, has to be abandoned. Generally, multiple slicing techniques

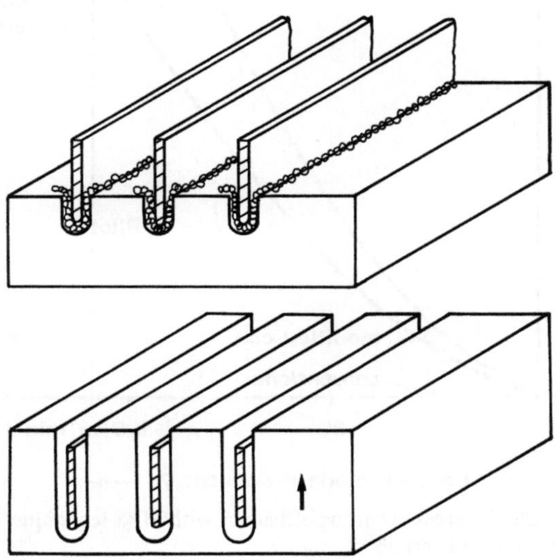

Fig. 26. Schematic view of multiple-blade slurry (MBS) slicing technique

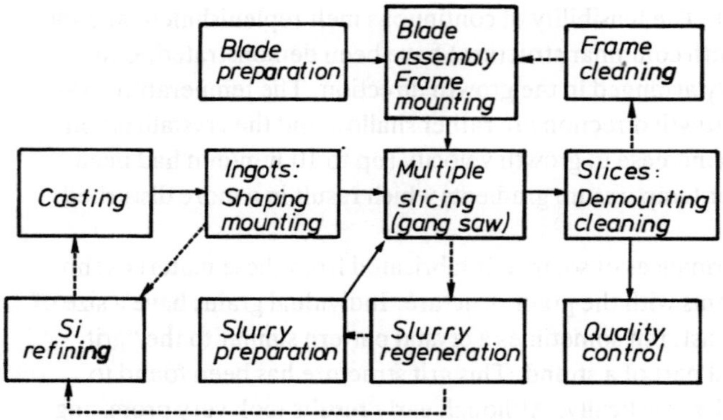

Fig. 27. Flow chart for cutting ingots (OD) and slicing with MBS technique[9]

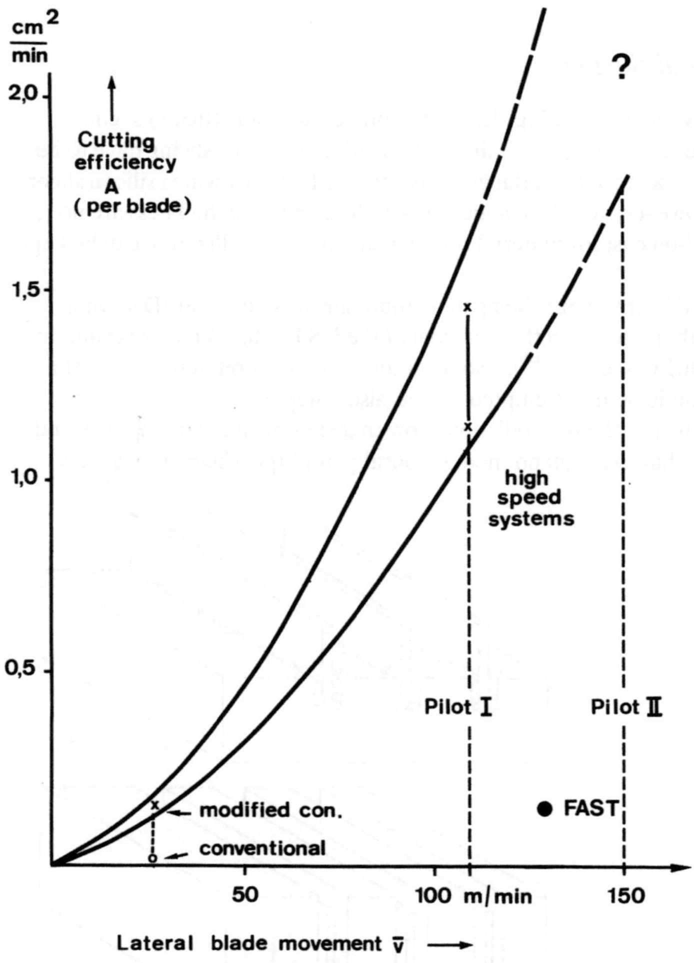

Fig. 28. Progress in cutting efficiency with MBS technique[9] for comparison, a value for FAST is inserted in this graph[113]

appear to be the only high-potential answer to this particular challenge. Several technologies are now emerging which show the important criteria to achieve the necessary cost reduction in the slicing process. The slicing technologies that are under investigation include multi-blade slurry (MBS) technique, multi-wire slurry (MWS) technique and a multi-wire technique with wires carrying an abrasive coating (fixed abrasive slicing technique – FAST).

The MBS-technique[105)] utilizes the cutting effect of steel blades in the presence of a slurry carrying abrasive particles (a schematic picture is given in Fig. 26). It combines low equipment and labor cost with high cutting efficiency. The basic idea of utilizing as much blades per unit as possible has led from the beginning to a heavy machinery concept. The frame weighing nearly one metric ton caused severe difficulties that have been tried to overcome by moving the ingot-bearing system. In another project[106)] the "harp" of blades is moved at much higher frequencies than reported and verified before. In terms of the mechanical concept of such an equipment, development and evaluation has been emphasized on pressure, blade length, stroke length and oscillation frequency[10))]. The slurry system had to be optimized with regard to viscosity and nature resp. concentration of abrasive material[108)]. Figure 27 gives a survey of the different areas of MBS-technique which have to be taken into account. The progress in cutting efficiency that has been reached up to now is shown in Fig. 28. A value of 2 cm^2/min/blade is expected during full operation of the pilot II system.

The MWS technique[109–111)], utilizes a single cutting wire wound onto a delivery spool. The wire is arranged in multiple loops around grooved drive and idle rollers and fixed to the take-up spool. A schematic draft of this technique is given in Fig. 29. The MWS

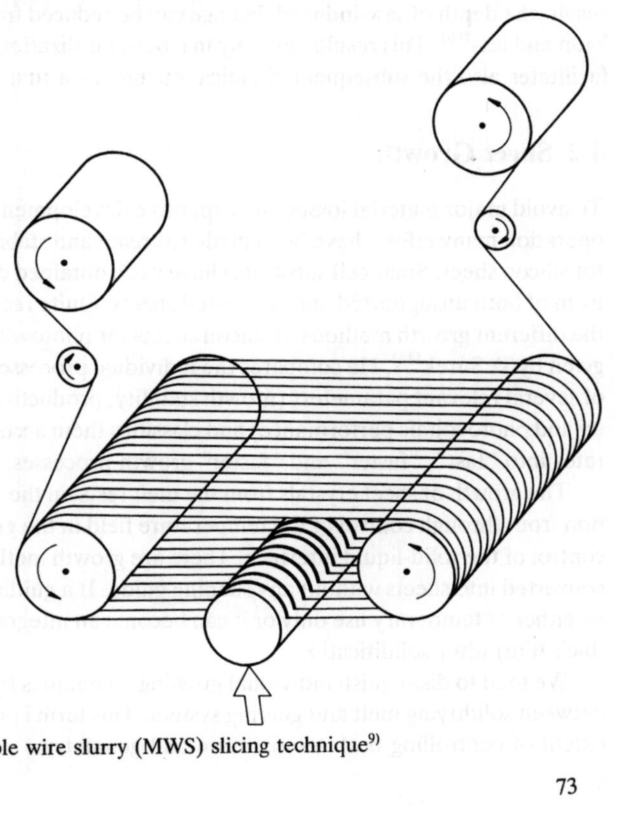

Fig. 29. Schematic view of multiple wire slurry (MWS) slicing technique[9)]

technique is capable of low kerf loss operation. Currently, up to 25 wafers per centimeter can be cut, i.e. this technique allows very efficient material utilization. The cutting efficiency is markedly lower than in MBS technique, due to considerably lower working pressures. The system is very complicated and presently requires far too much of the operator's attention to run it in a production environment. In addition, the operational conditions are not yet well understood thus leading to a low performance of the cutting process. Wire breakage and lifetime, slurry hose ruptures, bearings which easily fill with slurry etc. are major concerns.

Extensive tests have already proved that MWS techniques may offer, in general, a high economical potential but the present engineering problems have to be solved before it can be utilized in a cost-saving production line.

FAST is based on a "harp" of wires with fixed abrasive[122, 113)]. Simple mechanical handling and extremely low kerf loss are quoted as the major advantages of this special gang-saw method. Essential features of the set-up are light-weight wirehead capable of high lateral speed, sensitive feed mechanism, counter-balanced drive, crystal rocking assembly and vibrational isolation of the drive unit. An important requirement of FAST is that diamond impregnation of the cutting wire is uniform and does not impair its mechanical strength and tension. Impregnation of diamond into the bottom cutting edge helps to reduce kerf loss and adds to the accuracy of slicing. Cutting efficiency is not very high; economic analysis[113)], however, shows lower add-on cost (1980 US-$ 6.48/m^2 sheet) as compared with the other wafering techniques (MBS, MWS, ID; see[103)]).

Unconventional slicing techniques in common have one distinguished characteristic. In contrast to ID technique, the cutting process can be carried out more gently. As a result, the depth of saw-induced damage can be reduced from more than 20 μm to about 7 μm and less[106)]. This results not only in a better utilization of ingot material but facilitates also the subsequent chemical etching as a first step in solar-cell preparation.

4.2 Sheet Growth

To avoid major material losses and expensive development of a high-efficiency slicing operation many efforts have been made to create and stabilize direct growth processes for silicon sheet. Solar cell substrates have been obtained directly from the melt in the form of both unsupported and supported sheets. Quite recently, an excellent survey of the different growth methods of silicon sheets for photovoltaic applications has been given by T. Surek[115)]. He compares the individual processes with each other on the basis of several relevant parameters (growth stability, productivity, impurity effects, crystallinity and photovoltaic performance) and classifies them according to their linear growth rates into "fast", "faster" and "fastet" growth processes.

The growth of sheet crystals from the melt rests on the exact control of the crystallization front through control of the temperature field in the growth system and through control of the solid-liquid interface. There are growth methods where the silicon melt is converted into sheets without any shaping guide. If a guiding system is applied, this can be either of temporary use only or it can become an integral part of the "ribbon" (as a thick film) after solidification.

We tend to distinguish individual growing techniques by the degree of interaction between solidifying melt and guiding system. This term is strongly correlated with the extent of controlling both temperature and geometry.

Four categories of interactions can be named:

a) Very weak interactions (between the guiding system and the melt to be crystallized) exist when solid material is transferred (e.g. by zone melting) into silicon sheet (RTR, HRG, LASS, see Sect. 4.2.1).
b) Weak interactions occur when the guiding system is applied only to a certain part of the molten source (WEB, ESP, ICC, roller quenchingsee Sect. 4.2.2)
c) Strong interactions can be seen in cases where the melt is forced through a corset-like guiding system (Stepanov, EFG, CAST, see Sect. 4.2.3)
d) Very strong interactions occur when the guiding system remains in close contact even after solidification of the melt.

As a direct result of an interaction between the shaping guide and the liquid/solid silicon system different forms of the crystallization front can be described in correlation with the ribbon geometry. If the solid/liquid interface is oriented perpendicular to the crystal pulling direction, heat dissipation can occur only via the ribbon cross-section (RTR, EFG, CAST). In cases of a biased crystallization front (wedge type) heat extraction can work much more efficiently, thus providing substantially increased pulling rates. Exhaustive information on silicon sheet material is supplied by Brissot[116] and by a special issue of the Journal of Crystal Growth[117].

4.2.1 Very Weak Interactions

In the ribbon-to-ribbon (RTR) process[118, 119] no special shaping guide is used; the shape of the meniscus is inherently controlled by the geometry of the feed material based on the high surface tension of silicon. The principles of this process are schematically given in Fig. 30.

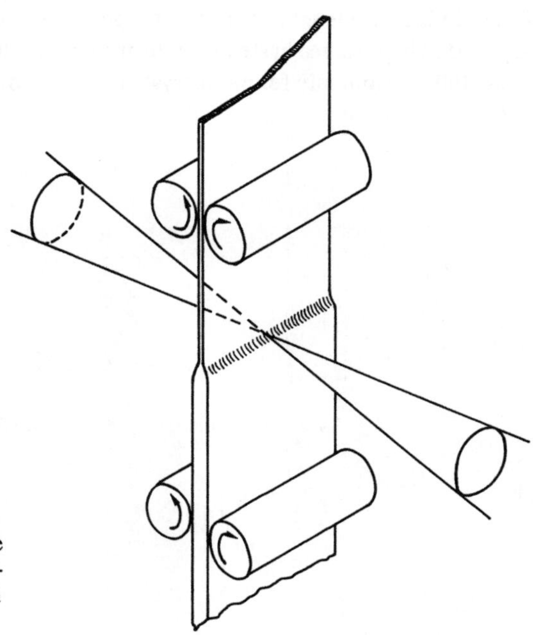

Fig. 30. Schematic graph of ribbon-to-ribbon (RTR) technology. Molten zone between feed stock and ribbon is generated by laser scanning or electron beam action

The RTR process offers possibilities of generating high-purity ribbons, due to the absence of any foreign guide material. Segregational effects are low in this case but evaporation may be an additional purification factor. A major problem, however, is the contamination of the microcrystalline feed ribbon by molybdenum during its CVD preparation on a substrate of this kind. Two different solutions are suggested here:
a) Removal of surface contaminations on the feed ribbon by etching or
b) Surface protection of the substrate by coating with e.g. Si_3N_4.

The small width of the molten zone and the steep thermal gradients correlated with it exert considerable mechanical stress on the solidified silicon and may lead to fractures. Special afterheating is applied to relieve the strain. The polycrystalline RTR-grown specimens exhibit high densities of dislocations (10^6–$10^7/cm^2$) and microprecipitates ($10^{13}/cm^3$)[120]. Gradual improvements of this process have resulted in ribbon grades that have led to solar cells from originally approx. 5% to the present maximum of 10.5% conversion efficiency (AM 1)[121]. H. Rodot[122] has proposed a method by which a (monocrystalline) ribbon is tangentially pulled from a stationary melting zone being generated (by an electron gun) in the peripheral area of a rotating polycrystalline rod. Presently, this process seems to be in a very early stage of testing.

In contrast to both methods mentioned which are characterized by narrow melting regions, Horizontal Ribbon Growth (HRG)[123–125] and Low-Angle Silicon Sheet (LSSA) growth[126] occur from a larger pool of molten silicon. In both cases, the ribbon is pulled off the meniscus of the melt under low-angle conditions. Over a larger area the surface is uniformly gas-cooled to create a well-sized undercooled area providing controllable steady-state conditions in terms of solid-liquid interface.

HRG is schematically described in Fig. 31. It has been improved – compared to its earlier version described by Bleil[127, 128] – in two essential points:
a) Gas cooling is used for the dissipation of the heat of solidification instead of a solid-heat extractor utilized in the original Bleil process.
b) The ratio L/t (L = length of growth wedge, t = thickness of the growing ribbon) is adjusted to high values in steady-state growth; this leads to very high linear growth rates (100–415 mm/min for monocrystalline growth, 850 mm/min for polycrystalline ribbons[125]).

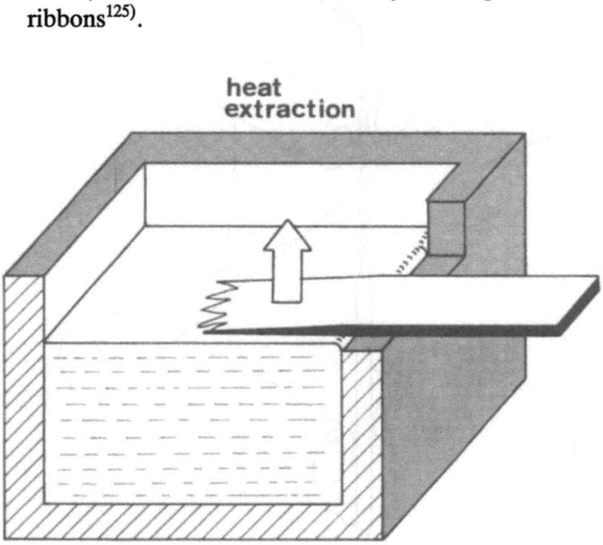

Fig. 31. Schematic graph of horizontal ribbon growth (HRG) resp. low-angle silicon sheet (LASS) growth. Angle between ribbon and surface of silicon melt ranges between 4° and 10°. Pulling rate influenced by length of wedge-shaped solidification front

In this relatively complex process, precise control of temperature profile and melt level is imperative. Up to now, the uniformity of the ribbon in terms of width and thickness still needs further improvement. Typical data of HRG and LASS-grown ribbons are given in[115, 125]. The quality of HRG-ribbons pulled at Nippon Silicon can be compared with that of CZ wafers. AM 1 conversion efficiency data in the range of 9–10% (without AR coating), as they have been measured there, are quite promising.

4.2.2 Weak Interactions

In the growth processes described below shaping guides are used which either interact with one of the two dominant ribbon surfaces or stabilize both edges.

In the case of Wacker's "Interface-Controlled Crystallization" (ICC)[129] crystallization takes place in the form of a coarse-structured multicrystalline silicon foil on a cooled ramp protected by a liquid film of molten slag. The latter supports the spreading of molten silicon as a thick film which is pulled off in tangential direction (see schematic drawing in Fig. 32). In a lab-type set-up ribbons have been made that were 800 mm long, 50 mm wide and 0.4 mm thick – at linear growth rates up to 120 mm/min. As the temperature conditions at larger dimensions are hardly affected by the pulling operation we can expect a considerable increase of growth rates when working in pilot systems. A real production unit for continuous ribbon growth by ICC may be designed in a way that is shown schematically in Fig. 33.

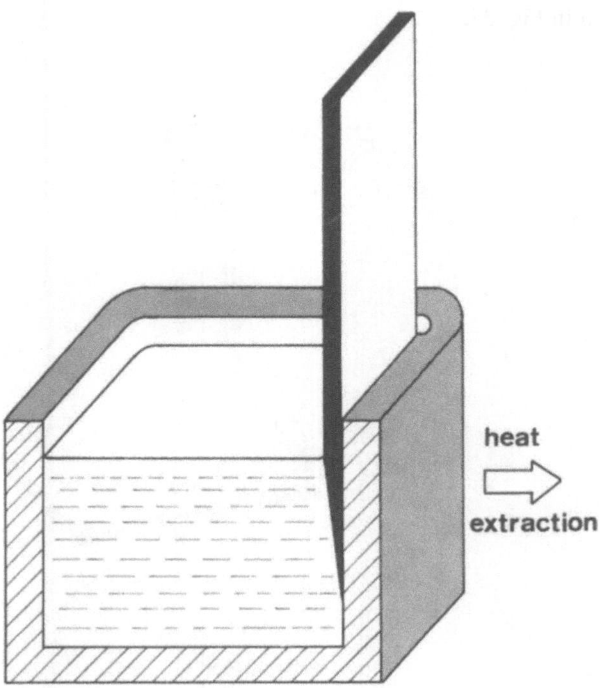

Fig. 32. Schematic graph of interface-controlled crystallization (ICC) method. The ribbon crystallizes at the cooled ramp which is covered by a liquid film of molten slag to avoid sticking of solidified silicon

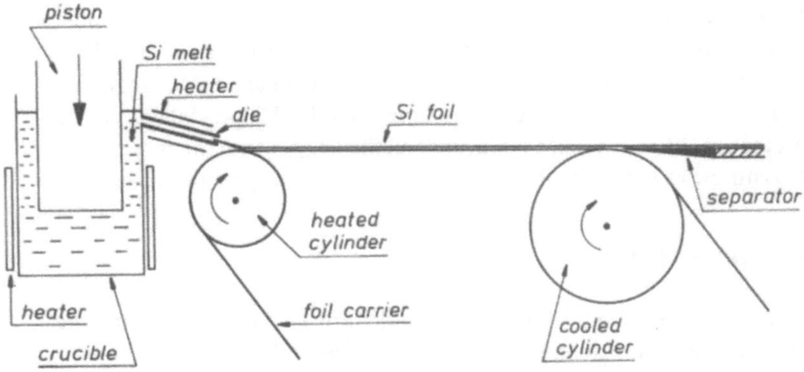

Fig. 33. Schematic view of a set-up for continuous production of ribbons by a modified ICC method

A rapid quenching equipment designed for producing amorphous metal foils at Tohoku University has been utilized to form polycrystalline silicon ribbons under extremely fast growth conditions[130–132]. In this so-called roller quenching method, molten silicon is ejected through a nozzle onto a very cold and fast rotating cylinder or between a pair of rotating rollers. The surface fo the cylinder or roller, respectively, is moving with a very high lateral velocity. It can be seen as a sort of fast moving stabilizing ramp that carries the solidified melt away. A schematic representation of this method is given in Fig. 34.

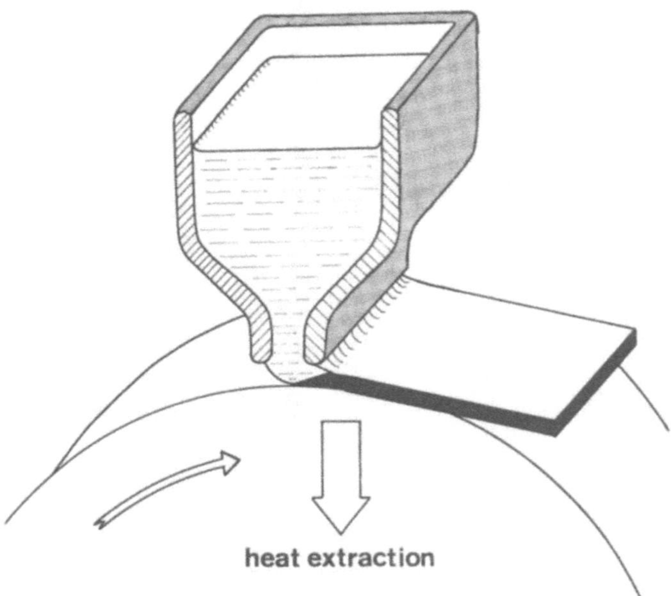

Fig. 34. Schematic graph of rapid quenching (roller quenching) method. A liquid silicon film is quenched in contact with the surface of a very fast rotating drum

The molten silicon is cooled very rapidly (quenched) to yield ribbons (1 to 50 mm wide, 20–200 μm thick). The extremely high growth rates of 10–40 m/s which have been obtained up to now, are very attractive. Some more items of interest are:

a) The apparent simplicity of this technique and
b) The well-established technology for producing amorphous materials in ribbon form.

The rapid quenching method by nature, lends itself to automation. Extensive efforts will be required yet to solve problems of small grain size, width and thickness control, mechanical stress, and uncontrolled contamination.

A guiding system stabilizing the edges of a ribbon is used in the edge-supported pulling (ESP)[133)] method and the dendritic web growth (WEB)[134, 135)]. Already in 1966, Tsivinsky et al. proposed a method using stabilizing edges to grow Ge ribbons[136)]. Two tungsten rods grooved along their axes were placed parallel to one another so that they emerged from the free surface. Due to wetting and capillary action, the melt rose up to the ends of the rods which served as a frame for the liquid film. During growth, the liquid/solid interface was held a little above the upper ends of the rods. ESP represents a method that utilizes a pair of quartz or graphite filaments, arranged parallel to each other, as nucleation aid. In the WEB process, the heterogeneous filaments are replaced by a pair of dendritic silicon needles being grown during the early stages of sophisticated seeding. In both cases, a thin Si film is formed in between and pulled off the silicon "pool" to form a web-like ribbon. The WEB/ESP method is schematically depicted in Fig. 35.

In the WEB process, besides EFG (see Section 4.2.3), the majority of developmental work in crystallization processes has been investigated. (Early studies already started in the late 1950's.) After four years of development within the LSSA Project, Westinghouse presently tries to reach the feasibility stage in terms of low-cost production[135, 137, 138)] and

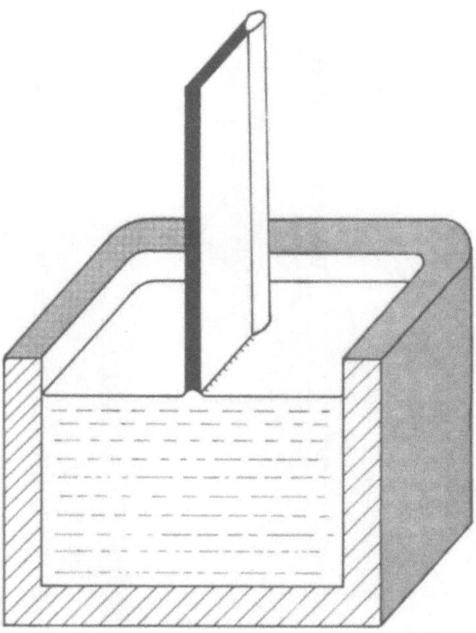

Fig. 35. Schematic graph of dendritic web growth (WEB) resp. edge-supported pulling (ESP) method. The edges are stabilized either by dendrites grown simultaneously with the silicon sheet or by graphite or quartz filaments fed through the melt to the crystallization front

to start an early production line. Under the aspects of growth stability, productivity, purity, crystallinity, and solar-cell testing, the WEB process is valued very highly[107] and, according to [95], is one of the most favored candidates to reach the DOE-LSSA goal of $ 0.70/W_p (in 1980 dollars).

In contrast, ESP has been investigated quite recently at SERI. This process can be controlled more easily in terms of thermal stability.

Ribbons exhibiting a coarse multicrystalline structure (0.2–2 mm thick, 30–50 mm wide, and 450 cm long) lead to solar cells of some 10% conversion efficiency (CE) AM 1 (13% in the case of seeded growth). Although experiments on a larger scale could not be evaluated yet, it can be stated that this "modified web process" offers advantages over other methods.

4.2.3 Strong Interactions

All methods to be discussed in this section are characterized by the existence of a film of molten silicon being shaped and limited by a die. The solid-liquid interface is formed here perpendicular to the ribbon surface and to the growth direction as well. Accordingly, growth rates are to be expected only in the range of 10–100 mm/min.

In principle, two systems of this category are discernible, namely the capillary-attractive and the capillary-depressive case (wetting/non-wetting). For the "edge-defined film-fed growth" (EFG)[139, 140] process and its modification, the capillary action-shaping technique (CAST)[141, 142], the first type of die material is used (Fig. 36). In the original

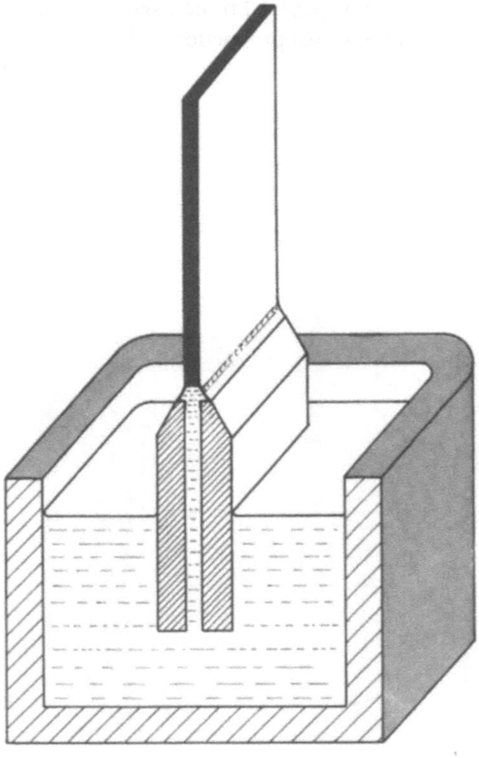

Fig. 36. Schematic graph of edge-defined film-fed growth (EFG) resp. capillary action shaping technique (CAST)

form of the Stepanov method[143, 144], low or missing capillary activity has been considered a prerequisite. Adequate transport of molten material, in such a case, has to be brought about by external forces. This has been the major reason to investigate the inverted Stepanov (IS) process[145, 146] where the hydrostatic pressure of the melt supply is utilized to accomplish what is necessary here (Fig. 37).

The EFG method, originally developed for manufacturing ribbons and tubes of sapphire, has been applied to silicon technology no earlier than 1972. However, it is the ribbon growth method that has most widely been studied. Intensive cooperation of experts from all disciplines has provided many information on most parameters that are important for mass production:

a) Growth stability can be sufficiently guaranteed by automatic control of temperature profile, meniscus height and ribbon width. This is possible with appropriate die design and the corresponding heating system.
b) The valuable results obtained in the meantime in automated continuous and multiple growth of ribbons 50–100 mm in width over periods of several days seem to bring the low-cost production within reach.
c) Contamination effects and crystallinity are mainly controllable, but there are principal problems due to $k_{eff} \approx 1$ and the unavoidable intensive contact between melt and die.

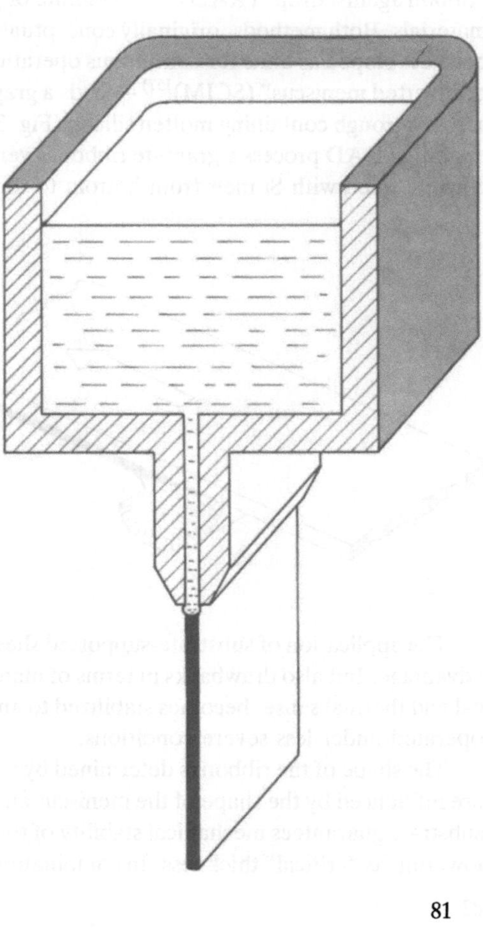

Fig. 37. Schematic graph of inverted Stepanov (IS) method

d) The relatively low conversion efficiency of solar cells made from such ribbons could be improved recently by additional impacts from the ambient atmosphere on the growth conditions[147].

For unknown reasons the EFG method has not yet seen the breakthrough in the US photovoltaic program that one would have expected. Due to problems related to points c) and d) some experts may see even better chances for WEB[103]. The modified EFG method is pursued less prejudiced in Japan[148–150]. There, this process has already reached a stage of commercial low-cost production of terrestrial solar arrays[151] although an average AM 1 conversion efficiency of no more than 8% has been reported.

The first results with CVD-coated dies are promising[152]. Sialon has been favored in this case on account of its potential purity and appreciable chemical resistance. Ribbons prepared on this basis so far are claimed to have reached η-values up to 11.8% (PVCE, AM 1).

4.2.4 Very Strong Interactions

In this category we find substrate-bound growth of silicon sheet (thick films) from the melt. In the best-known processes of this kind, "silicon on ceramics" (SOC)[153, 154] and "ribbon against drop" (RAD)[155, 156] mullite or graphite, respectively, are the supporting materials. Both methods, originally conceptualized as batch processes, meanwhile have been developed as units for continuous operation. SOC – in this version "silicon coating by inverted meniscus" (SCIM)[157] – needs a graphitized ceramic substrate which is moved across a trough containing molten silicon (Fig. 38). In the case of the continuously operating RAD process a graphite ribbon is vertically pulled through a bottom-slotted crucible filled with Si melt from bottom to top[158] (see Fig. 39).

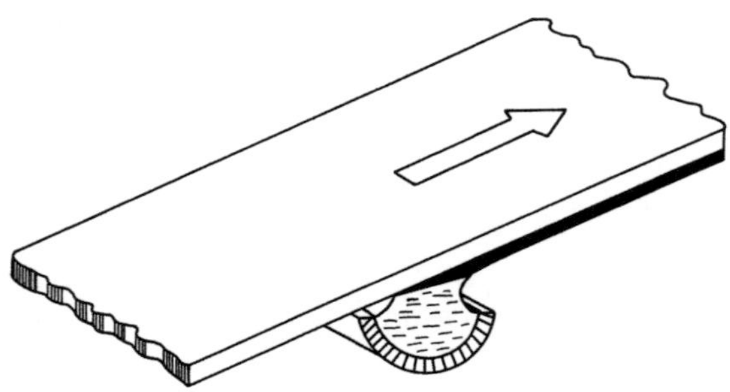

Fig. 38. Schematic graph of silicon coating by inverted meniscus (SCIM) method – as a further development of the original silicon on ceramics (SOC) method

The application of substrate-supported sheet growth techniques offers economical advantages but also drawbacks in terms of material quality. The meniscus, in a mechanical and thermal sense, becomes stabilized to an extent that thermal control can be operated under less severe conditions.

The shape of the ribbon is determined by substrate geometry. Width and thickness are influenced by the shape of the meniscus and the lateral substrate movement. The substrate guarantees mechanical stability of the silicon sheet and offers the chance of lowering its "critical" thickness. In combination with an improved form of heat extraction

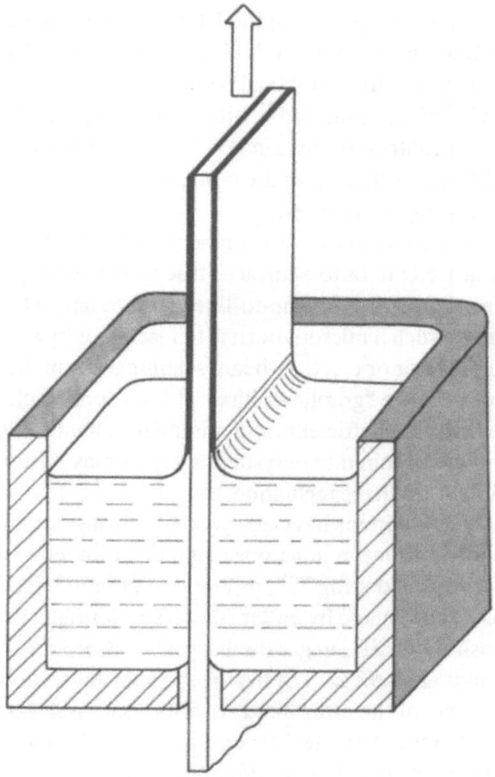

Fig. 39. Schematic graph of ribbon-against-drop (RAD) pulling method. Graphite foil is fed through bottom of melt crucible

across the interface this leads to higher effective pulling rates as compared with the methods described in Sect. 4.2.3.

By the direct contact of melt and substrate some difficulties arise which are similar to those observed during solidification in a crucible (see Sect. 4.1.1). For this reason the substrates have to be resistant to molten silicon and show thermal expansion properties comparable to crystalline silicon. Sheet quality in terms of impurity levels and grain structure leads to PVCE AM 1 values (average) of 9 or 7% for mullite and graphite substrates, respectively.

5 Vapor-Grown Films

Utilization of gas-phase reactions in deposition techniques for terrestrial photovoltaics is limited by two factors:

a) Film thickness has to be kept at small values for economical reasons
b) Polycrystalline (CVD) material of small grain size, in particular, has never displayed adequate conversion efficiency

Consequently, either monocrystalline (epitaxial) or amorphous thin films are the only technical solution here.

In the early stage of the ERDA project, solutions have already been offered to make rather impure (polycrystalline) silicon material a workable material by vapor deposition of a "protective" epitaxial layer[159]. Experiments, have been continued on this basis at RCA[160] and Hitachi[161] with interesting results.

In contrast to this kind of "thin-film finishing" of more or less randomly oriented silicon solidified from the molten state, a very different way to utilize thin layer growth is increasingly discussed.

Based on a completely amorphous substrate material (e.g. glasses) vapor deposition is carried out onto a surface structure consisting of parallel grooves to enforce orientational growth. The line-to-line distance has to be very small and the technical solution to create such a micropattern is to inscribe it by a stencil, to etch it by microlithography or to apply laser or electron-beam scanning. From the first mentioned version the method has got its name "grapho-epitaxy"[162]. As soon as cheap material in terms of adjusted thermal expansion coefficients and minimized chemical interaction will be found, the direct growth of thin monocrystalline layers may be a true competitor among substrate-based ribbon or sheet techniques.

Expectations have also increased considerably in the field of amorphous silicon. Shortly after p-n junctions could be prepared in this exotic type of semiconductor by hydrogen "doping"[163], its low-cost potential has been highly stressed. Technically, films of a-Si are made by either silicon sputtering in a hydrogen atmosphere[164] or by decomposition of SiH_4 in gas discharges[165]. More recently, improved annihilation of unwanted bonding states in a-Si has been reported to occur by fluorine (SiF_4)[166].

One of the often cited advantages of a-Si is the practically complete absorption of light in the characteristic energy range within a few microns only (meaning small amounts of silicon, in terms of mass per square unit).

However, some factors being of major importance for the terrestrial photovoltaic application like PVCE data (at present between approx. 3 and 6%), homogeneity, reproducibility and long-term stability are not sufficiently known yet. Work is going on at many places to further clarify the problems as indicated (see e.g.[167]). Due to the complexity of this matter, a more thorough treatment would be beyond the scope of this article.

6 Material Characterization

6.1 Impurities. Analysis and General Characterization

The upper concentration limits for electrically active elements (dopants, impurities) in terrestrial solar-grade silicon lie in the region of 10^{-5} to 10^{-7} (Sect. 1.3, Fig. 4).

This concentration range is still detectable by modern instrumental methods of chemical analysis. A survey of atomic absorption (AA), atomic emission spectroscopy (AES), X-ray fluorescence, neutron activation and spark source mass spectrometry is given by Millet[168]. We distinguish between two types of methods:
Both AA and AES require samples prepared as solutions. For the second type, solid (bulk) samples can be used without any chemical preparation.

The signal obtained from most analytical instruments will not be related to element concentrations in a directly calculable manner and all instrumental measurements have to be calibrated. This is a simple matter for solution techniques as absolute artificial standards can be prepared. The signal from an impurity element can be modified by interactions with the major components (matrix) or even other trace impurities. The advantage of the solution techniques is based on their relative independence from such interferences. A limitation of all solution techniques is the background introduced by contamination. Some of the sources may be impurities of reagents, dust in the atmosphere or surface contamination of both apparatus and sample[169)].

In the second group, spark-source mass spectrometry (SSMS) seems to be the most useful method for analyzing TSG-Si. Spark-source mass spectrograph is the only analytical instrument capable of supplying information on practically all elements simultaneously with a sensitivity of 10^{-8}. It has been shown that measurements by SSMS can generally be relied upon with a factor of two, provided careful calibration is performed[59)].

This, however, needs comparative analysis based on standard specimens of similar composition, a characteristic limitation of all instrumental techniques using solid samples.

Impurities in silicon may be expected to affect solar-cell properties in a variety of ways. At high impurity concentrations, crystal growth can be perturbed by inclusions, precipitates[93)] or defects that may finally lead to structural breakdown. Semiconductor properties may be influenced by electrically active impurity centers which reduce the minority carrier diffusion length either by increased recombination or by scattering – induced mobility loss. Additional possible impurity mechanisms may include interface degradation and series or shunt resistance effects as well as other junction defects.

Tremendous efforts have to be made to correlate the concentration of impurities with microstructural defects, electrical properties and solar-cell characteristics. Such correlations are necessary for establishing the concentrations that can be tolerated in the silicon

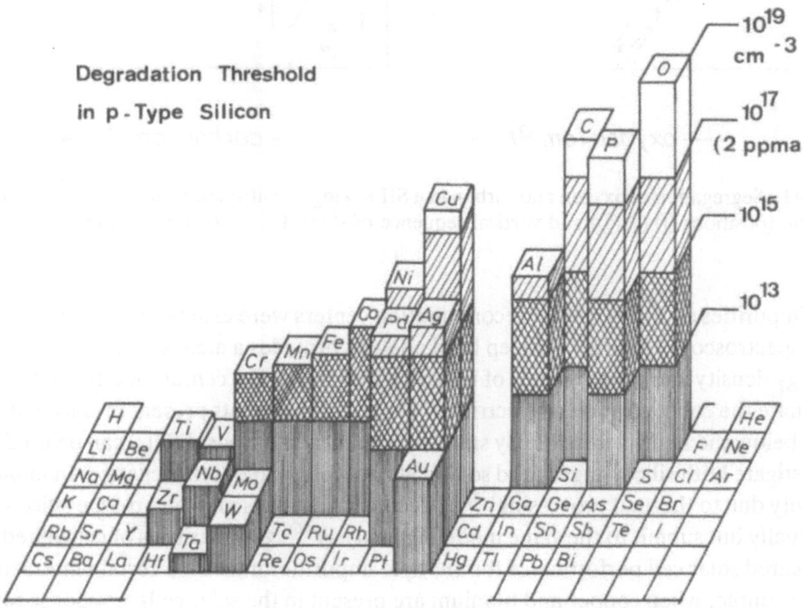

Fig. 40. Threshold values for solar cell degradation of different impurities in Si[10)]

crystals without degrading the solar-cell performance (Fig. 40). Very remarkable contributions to this subject have been made in the USA[10, 170] focussing so far on the detrimental influence of various transition metals on standard solar-cell performance. In Germany efforts have been concentrated on the major impurities in metallurgical-grade silicon. For instance, main group metals like Mg and Ca, of less significant interest in the field of conventional silicon research up to now, have been analyzed with regard to solubility and segregation coefficient data[171]. To understand the specific behavior of certain elements on their way through different thermochemical processing steps, additional attention is focussed on non-metals like the chalcogens S, Se and Te[172] or carbon and oxygen. Carbon and oxygen concentrations were measured by infrared absorption. A marked segregation is found for these slowly diffusing light elements[173] as shown in Fig. 41.

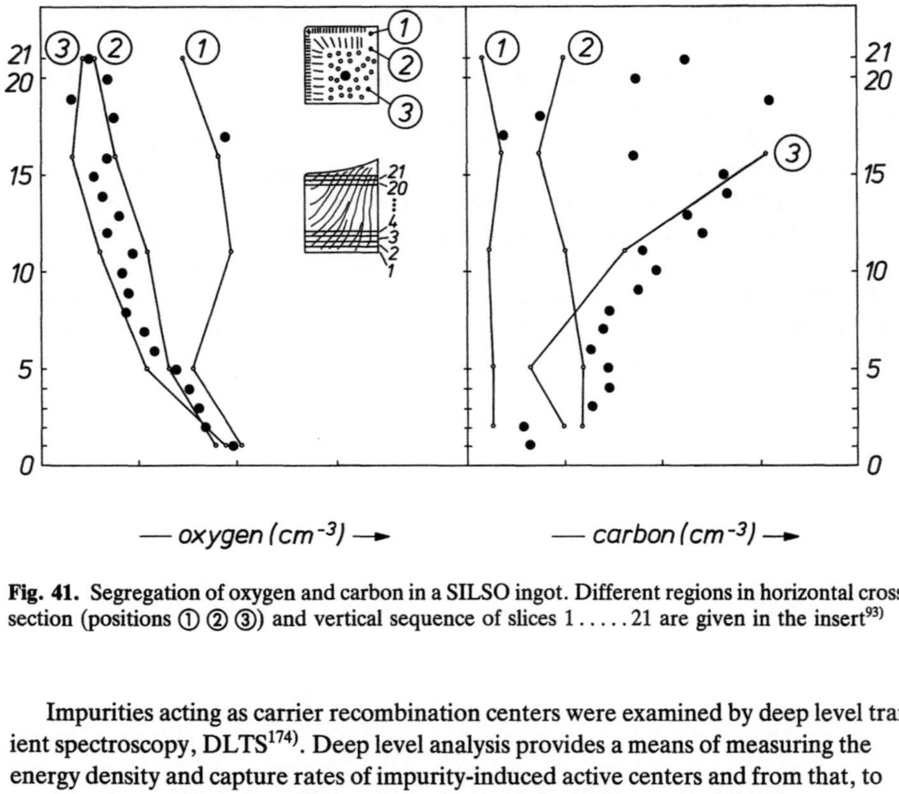

Fig. 41. Segregation of oxygen and carbon in a SILSO ingot. Different regions in horizontal cross-section (positions ① ② ③) and vertical sequence of slices 1 . . . . . 21 are given in the insert[93]

Impurities acting as carrier recombination centers were examined by deep level transient spectroscopy, DLTS[174]. Deep level analysis provides a means of measuring the energy density and capture rates of impurity-induced active centers and from that, to evaluate the concentration of electrically active contaminants present in silicon at levels well below the limits accessible by standard chemical methods. DLTS can be used to investigate both silicon ingots and solar cells, as well as to observe changes in impurity activity due to thermal processing[175]. Effects of impurities in n- and p-type silicon differ gradually but submit to the same modeling analysis[176]. Comparisons of calculated and measured solar cell performance for multiple impurities indicate a certain interaction. For example, when copper and titanium are present in the solar cells, copper seems to mitigate the unfavorable effects of titanium[177].

6.2 Photovoltaic Characterization

The photovoltaic performance of solar silicon is markedly influenced by the starting properties of the sheet material. There are mainly two situations which are encountered here:

- High content of unintentional impurities (zero-dimensional lattice defects) – either insufficiently removed during purification or introduced during sheet formation,
- 1-to 3-dimensional crystal defects are observed in both mono- and multicrystalline material.

Impurities and defects generally increase the saturation current density J_o thus lowering the open circuit voltage, V_{oc}. The short circuit current, J_{sc}, is reduced due to low minority carrier diffusion length, L_{eff}, and recombination at defect sites. In general, J_{sc} is a function of the minority carrier lifetime within individual crystallites, V_{oc} is mainly dependent on the lifetime situation near the depletion region. The fill-factor, FF, is reduced by a high J_o and by the presence of non-ideal currents. Unsuited surfaces and grain boundaries lead to excessive resistance or shortening of the junction. Impurities and defects must not have a direct harmful effect; they also can reduce cell performance by interaction with solar cell processing methods or materials used herein.

The photovoltaic (PV) parameters η, V_{oc}, I_{sc}, FF, and the diode characteristics are the main tools here by which the influence of impurities and defects can be investigated. These forms of direct PV evaluation are supplemented by measurements of minority carrier diffusion length L_{eff}, of spectral response, and of the dark diode characteristics.

On such a basis various papers have been dealing with thorough investigations on most different sheet materials[97, 178–180]. Other publications were concentrating more specifically on EFG ribbons[181], CAST ribbons[182], WEB ribbons[183], RAD ribbons[156] or SILSO slices[184, 185], respectively – to name just a few. Efforts have been concentrated on studies of the influence of crystal size in a multigrain structure on several PV parameters, particularly on conversion efficiency (CE)[186–189]. The fact that for diameters > approx. 1 mm, CE is constant was corroborated repeatedly. In this case L_{eff} is not yet affected. Since grain boundaries (if they are electrically active) act as perfect sinks for minority carriers, the effective diffusion length is reduced with decreasing grain size even if the intergranular carrier lifetime remains constant. "Grit structure" is a highly degraded grain pattern with additional accumulation of impurities[93, 190]. In a region of this kind the values of V_{oc}, I_{sc}, FF, and L_{eff} decrease drastically (Fig. 42). The solar cell no longer shows diode characteristics and the current-voltage relationship is linear. In other words, this area does not generate a photocurrent and acts as a shunt resistor thus shortening the p-n junction of the remaining good cell area[191]. Further investigations also exhibited high concentrations of C and O in those critical regions[190].

As already mentioned, L_{eff} is the chief parameter in the control of CE. The most detrimental influence on it has been observed in the case of high-energy grain boundaries and dislocations. Therefore, major targets in further development of multicrystalline silicon are: enlargement of the average grain size and reduction of the dislocation density.

The direct impact of specific (unwanted) impurities on L_{eff} could be studied so far in a few practical cases only. In the RTR process molybdenum is transferred from the Mo-substrates via the polycrystalline CVD layer to the float-zoned ribbon (partially due to unfavorable segregational conditions)[197]. By using graphite material either as substrates

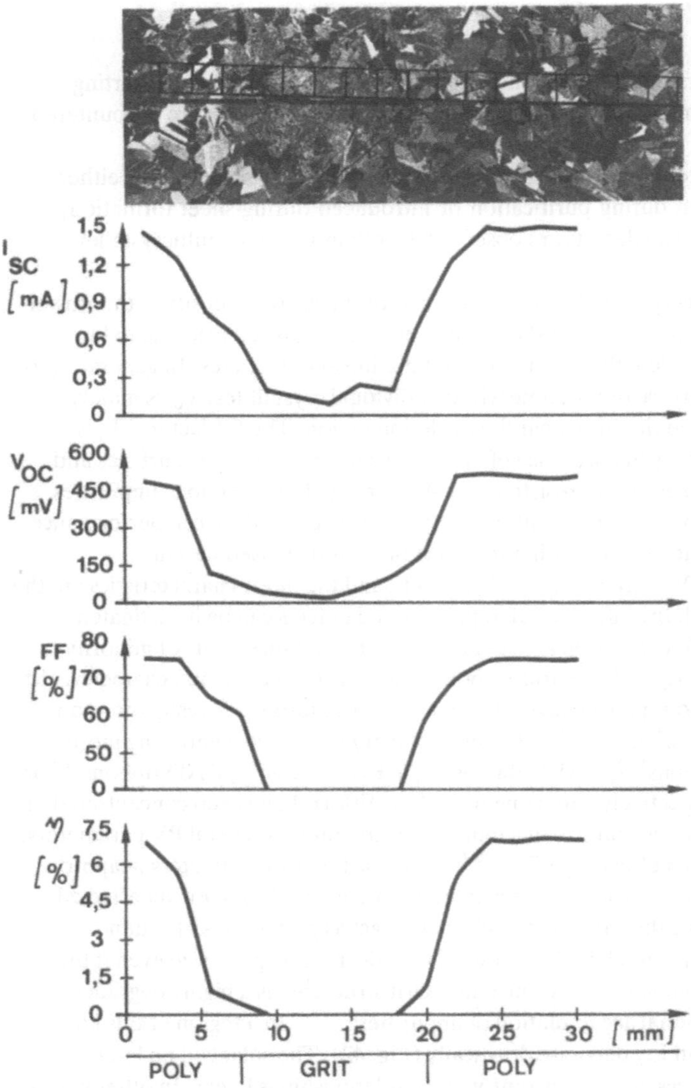

Fig. 42. Influence of unfavorable crystalline structure on I_{sc}, V_{oc}, FF, η. Distribution of photovoltaic parameters have been evaluated by a diode array[191)]

and dies or as molds, the resulting high C (SiC) concentration in the products may cause appreciable degradation of PV CE[156, 181, 182, 190)].

In the case of upgraded metallurgical-grade (UMG) silicon which has been processed further by the HEM method the following test results have been reported: Compared to standard MG-Si (>2.1%)[192)], CE values have been found for high-quality MG-Si in the range of 3.5–7.2%, depending on impurity levels[193)].

Low-cost Battelle pelletized silicon could be used without difficulty to grow silicon "webs"[134)]. Diagnostic solar cells exhibited average efficiencies of 12.8% AM 1 (calculated from a CE AM 1 value of 9.0% without AR-coating).

To investigate grain boundary influences on intentionally contaminated samples, ingots of partially refined MG-Si have been made[93]. This material shows a high, inhomogeneous impurity content (resistivity about 0.1 Ohmcm, p-type) and high concentrations of precipitates within the grains. Due to high impurity levels, CE values of approx. 3.5% AM 1 and data for $L_{eff} > 5$ μm were observed[191]. L_{eff} data for standard-type SILSO range around 75 μm.

In other cases where EG-Si has not been used as base material, either conventional crystallization methods were applied or the sheet had been "finished" by (vapor) epitaxy. In both cases the direct influence of impurities could not be determined.

6.3 Topographic Evaluation

For developing a better insight into the synergistic effects of the various forms of crystal defects in silicon sheet, space-resolved measurement techniques, mainly in the microscopic range, can be a very valuable tool. Most of the crystal defects are spotted first by preferential etching techniques in combination with different types of optical microscopes[173, 184]. Inhomogeneities in the distribution of impurities may be detected by AES (Auger emission spectroscopy), SIMS (secondary ion mass spectrometry), both in the scanning modification, and NAA (neutron activation analysis) – coupled autoradiography[93]. These methods have been supplemented by IR microscopy[184].

Photovoltaic evaluation of similar character is preferably carried out by EBIC (electron beam-induced current)[185, 196] or LBIC (light beam-induced current)[197, 198]. In the latter case, small-beam diameter lasers (He, Ne) have been operated at wavelengths of 0.6328 μm or 1.15 μm, repsectively[199]. The strongest effects can be expected here by utilizing a specially tuned GaAs diode emitting 0.910 μm where the spectral sensibility of Si solar cells is highest[194]. For EBIC and LBIC measurements, the presence of a collecting barrier and ohmic contacts are needed. In order to prevent any alteration during preparation through necessary heat treatments, an electrolytic junction and a liquid eutectic (Ga, In) contact can be formed at room temperature. By these means it is possible to study the influences of each step in solar-cell processing. This method has successfully been applied to multicrystalline silicon specimens[200].

Comparative measurements between defect distribution and PV characteristics were made on most different non-monocrystalline materials, mostly based on EBIC: EFG ribbons[181, 201, 202], CAST ribbons[203], RTR ribbons[118], RAD ribbons[200], and various samples from cast ingots[93, 194, 198, 204, 205].

Kazmerski has reported grain boundary segregation phenomena[86, 206]. By the application of scanning versions of AES and SIMS to virginal grain boundaries (as-cleaved in situ) two types of inhomogeneities could be detected. The first one is characterized by bulk to grain boundary segregation, the second one occurs through segregational effects within the grain boundary system due to different states of interface energy. The latter case can also be studied by the different electrical behavior of such areas. Typical high-energy grain boundaries show current degradation in EBIC by 20–30% whereas twin boundaries have practically no influence (0–10%) on this figure. Precipitates in the bulk area, most remarkably, degrade the current to the same extent as high-energy grain boundaries.

The electrical properties of selected grain boundary regions have been investigated using high-resolution SAM and EBIC techniques[207]. Measurements showed that the electrical activity of the grain boundaries differs considerably in a given sample. This is attributed to accumulations of impurities in this region. In another contribution to this subject a combined quantitative EBIC and ion microprobe analysis has been applied to EFG ribbons[201]. A direct comparison of the spatial distribution of impurities and crystal defects is given in[93]. However, up to now an ideal combination where the three different factors, namely impurities, lattice defects and PV properties, are brought into sufficient spatial correlation on one and the same sample is still missing. To reach this multidiagnostic level on a broad basis has to be one of the predominant targets in PV material science. Not till then PV experts will be in a position to master the adjustment of material properties to solar-cell processing and vice versa on the basis of optimized economical conditions (see also Sect. 6.4.2).

Subsequently, some representative examples related to multicrystalline silicon prepared by a mold-casting technique (SILSO) are given. To meet the fairly low detection limits of some analytical techniques applied here, a few ingots were cast from rather impure partially refined MG-Si. In all other cases, standard SILSO (1–5 ohm · cm, p-type) was used. The different analytical techniques have already been extensively described elsewhere[173, 184, 191, 193, 194].

Figure 43 a–e exhibits different micrographic appearances of carbon-contaminated SILSO samples. They indicate two important peculiarities: Bulk precipitation and a negligible correlation with the dislocation pattern (Fig. 43 c) – besides expectable grain boundary phenomena (Fig. 43 a, d). SiC precipitates were also detected in RAD ribbons by comparison of TEM[208] with ^{14}C autoradiography[203]. In attempting to clarify the nature of the so-called grit structure, topographic IR analysis revealed carbon peak values in correlation with its typical pattern (Fig. 44 a, b). An autoradiogram of a 50×50 cm^2 SILSO slice (Fig. 45 a) is shown in Fig. 45 b. In addition to typical grain-boundary segregation one observes impurity pile-up in the peripheral chill zone, in the regions of "grit structure", and in the central part of ultimate solidification. In areas of high impurity concentrations elements such as Sb, Ca, Eu, Ti, Co, and Fe have been found in particular. The degree of enrichment, of course, varies due to the different segregation and diffusion coefficients.

Concentration of electrically active impurities in the close neighborhood of SiC precipitates has also been detected in EFG ribbons[201] (see also[203] for CAST ribbons)

A higher magnification of the transcrystallization zone (Fig. 46 a) demonstrates that grain boundaries act as impurity captors and thus decrease the impurity content in individual (bulk) grains. The impurities are distributed very inhomogeneously along the grain boundary. A correlation has been found between local enrichment of impurities and edge-dislocation density in the grain-boundary region of a silicon bicrystal by aid of TEM[210] (see Fig. 50). Sometimes, grains which are bordered exclusively by coherent or twin boundaries have a very low impurity content. A close-up view of a grain boundary-free region indicates that the "haze" within a grain is due to a high concentration of precipitates (Fig. 46 b). EBIC and LBIC photographs of the multicrystalline structure of SILSO wafers are compared in Figs. 47 a, b.

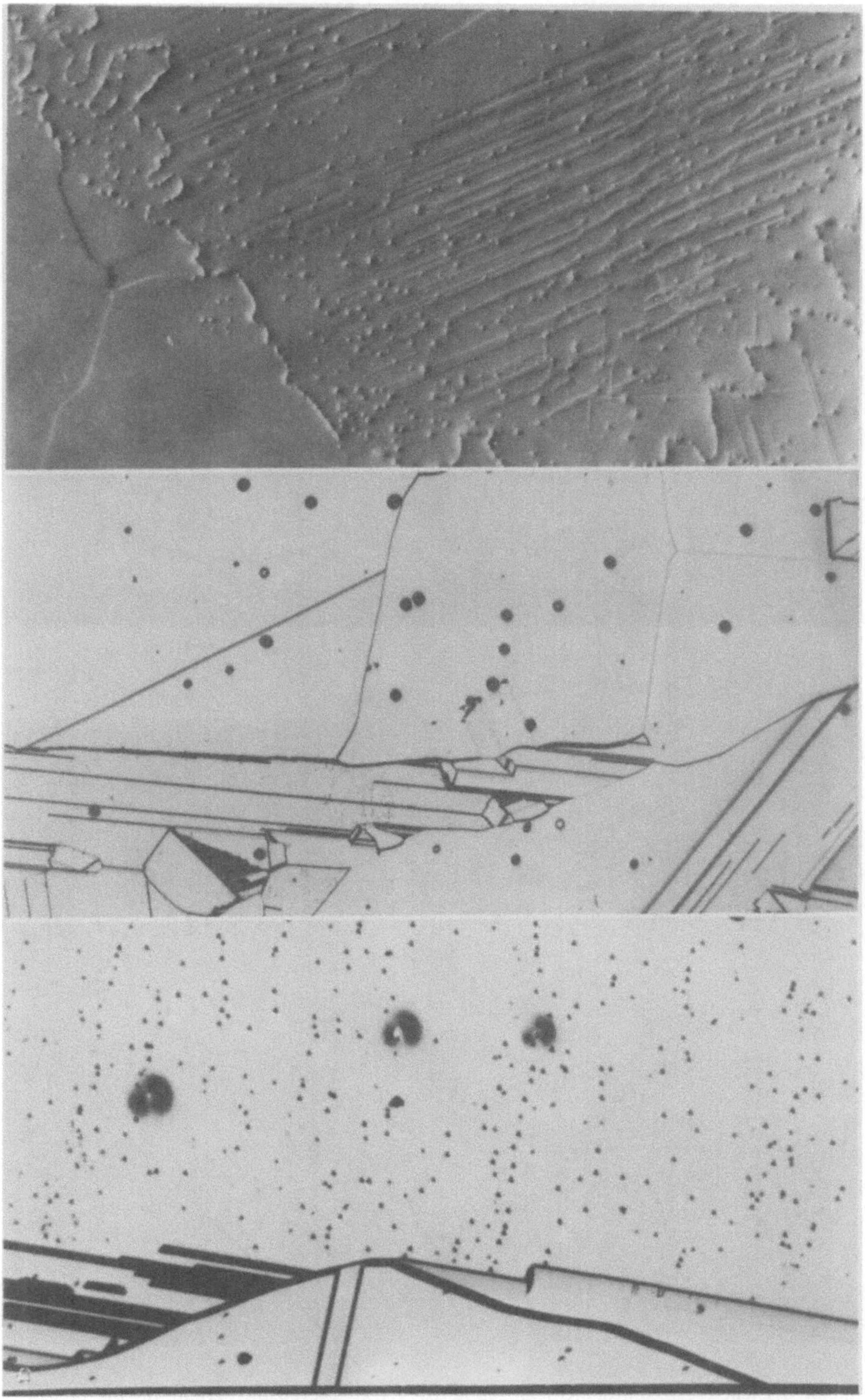

a

b

c

Fig. 43 a–c. Optical micrographs of SILSO contaminated by carbon. **a** mechanically polished surface (35 ×), **b** preferentially etched surface, CrO_3/HF (35 ×), **c** same surface (150 ×).

d

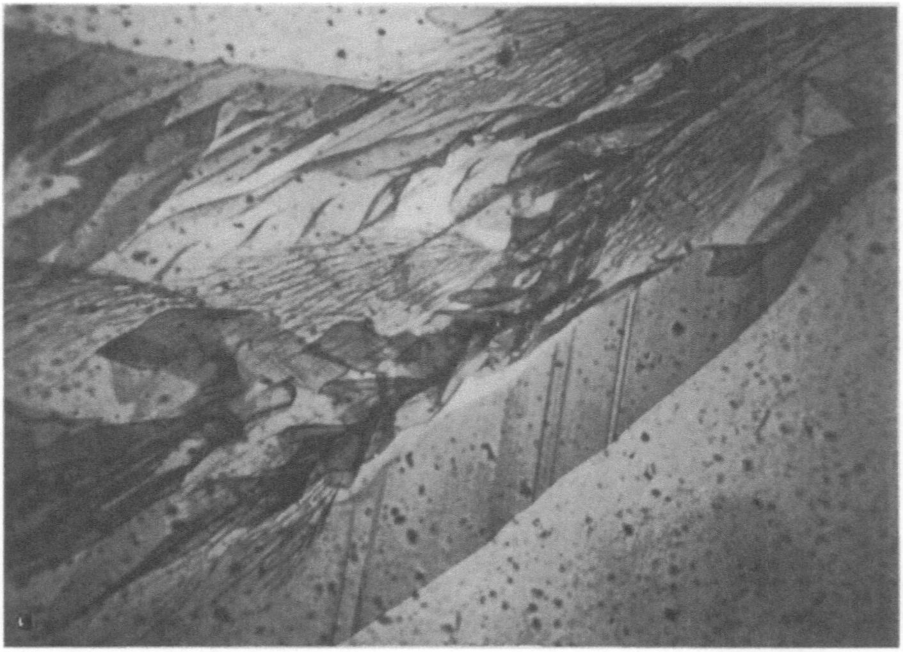

e

Fig. 43. d Etched surface with SiC segregation (200 ×); **e** infrared micrograph with SiC precipitates in grain boundaries and bulk region (200 ×) – Courtesy of AEG-Telefunken[184]

a

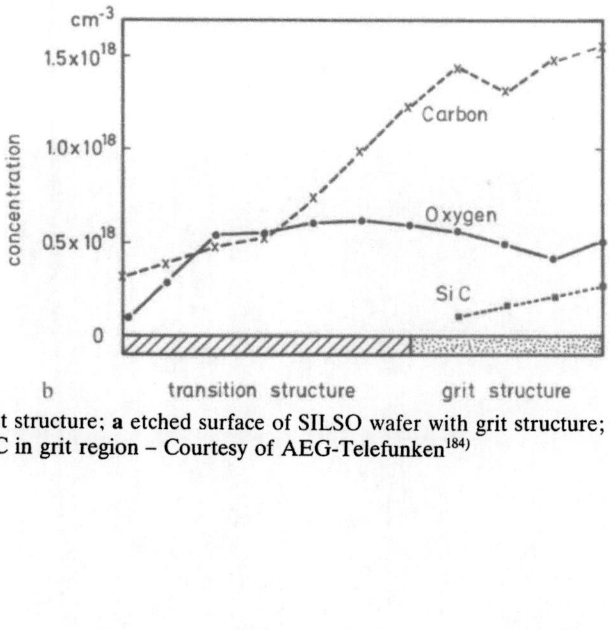

b

Fig. 44 a, b. Appearance of grit structure; **a** etched surface of SILSO wafer with grit structure; **b** distribution of C, O, and SiC in grit region – Courtesy of AEG-Telefunken[184)]

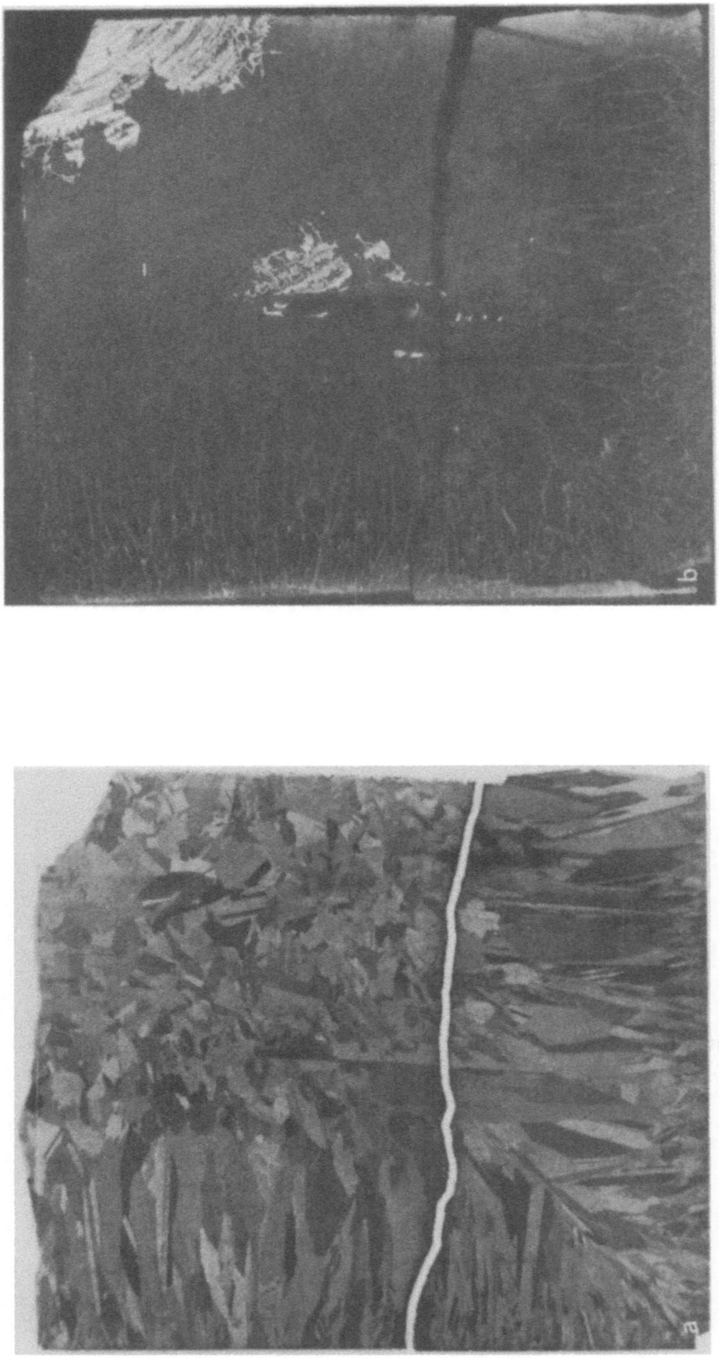

Fig. 45 a, b. Autoradiography (**b**) of a SILSO wafer (**a**) (etched sample); magnification 4 ×. The ingot has been prepared from partially refined metallurgical-grade silicon. High impurity concentrations appear as bright pattern[93])

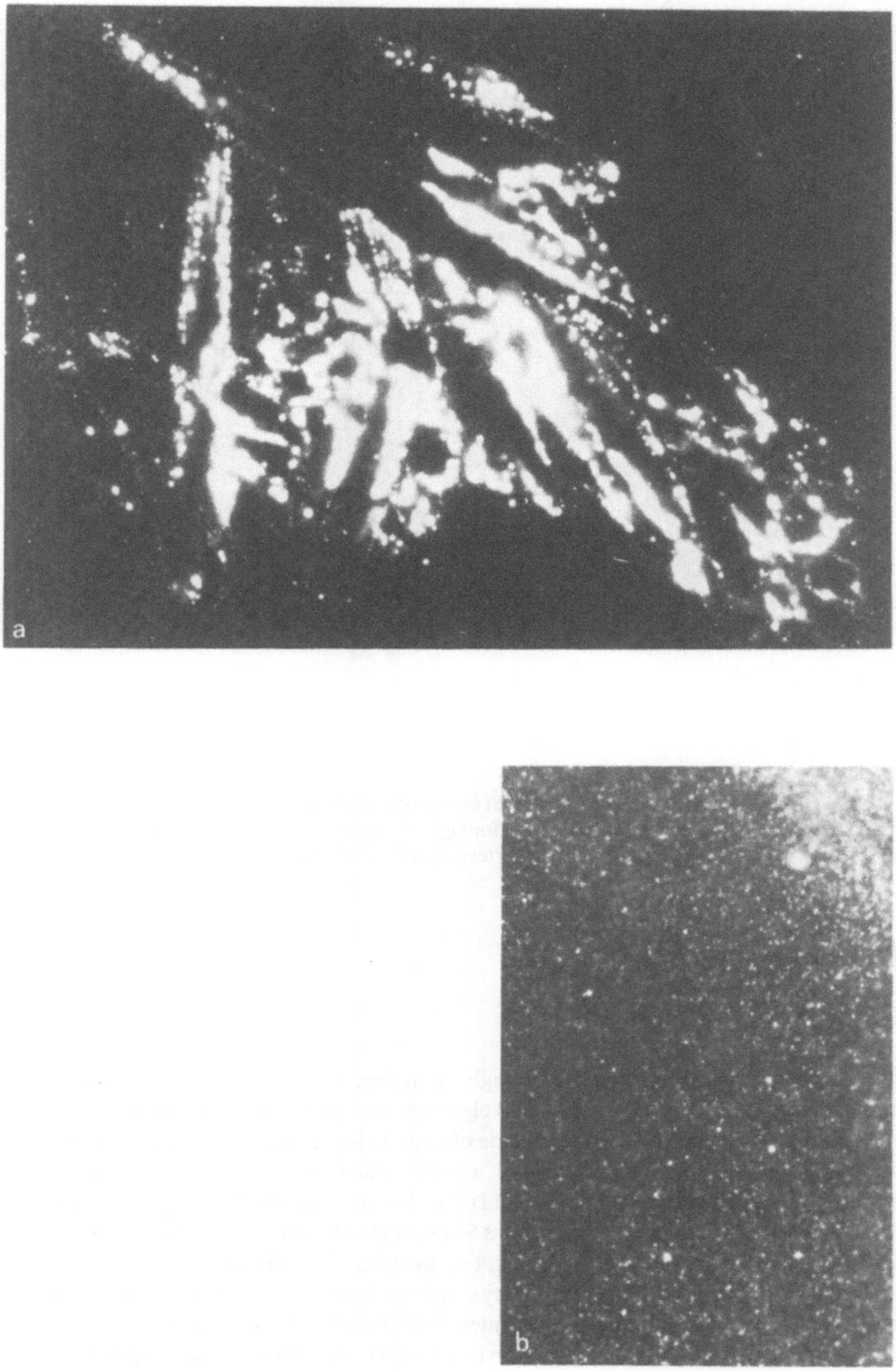

Fig. 46 a, b. Close-up view of autoradiography in Fig. 45. **a** Autoradiography of transcrystallization zone; magnification 2,5 ×; **b** Autoradiography of bulk precipitates; magnification 10 ×[93)]

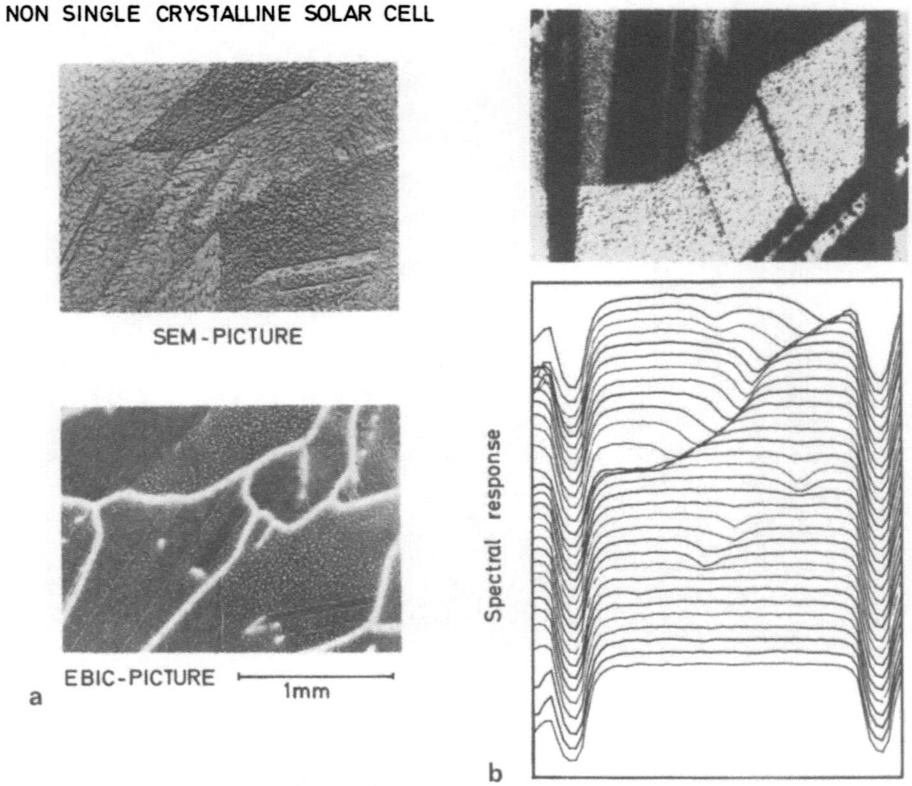

Fig. 47 a, b. EBIC and spectral photoresponse of multigrain solar cell. **a** EBIC pattern compared with SEM pattern. Bright regions are indicative for high recombination rates; **b** Spectral response of solar cell at a wave length of 920 nm – Courtesy of AEG-Telefunken[194)]

6.4 Grain Boundaries

6.4.1 Structural Aspects

Different types of bonding systems for a single atom already exist in the intrinsic cases of silicon lattice defects. One or two unbound electrons per atom ("dangling bonds") are typical of atoms being placed at the frontside of extra half planes forming edge dislocations. Other bonding states like trivalent Si^+ or pentavalent Si^-, as modifications of the normal sp^3 hybrid, have particularly been dicussed in connection with amorphous silicon, a-Si. The first-mentioned type of anomalous Si atoms (broken or dangling bonds) provides a maximum strain relief but is highest, too, in electrical interaction (traps). The other category of silicon atoms is completely saturated in terms of covalent bonding. Its contribution to lattice strain, however, is much stronger in a comparative sense. This leads to typical dualistic situations for any type of defect – from zero- to three-dimensional ones – where mostly unknown additional parameters decide on prevalence.

In Fig. 48 such counterparts within a given defect system are represented as the broken – and the restored-bond variety of a typical edge dislocation in the diamond

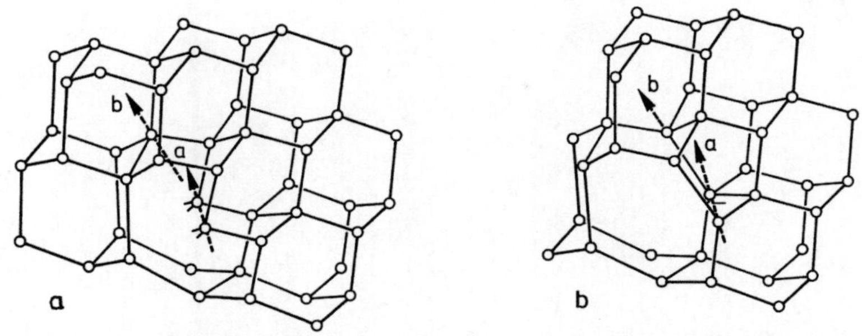

Fig. 48 a, b. Edge dislocation, {100} glide plane (a = axis, b = Burgers vector). **a** With dangling bonds; **b** with restored bonds. From Hornstra[211]

lattice[211]. In the restored bond system one encounters the co-existence of pentagonal and heptagonal rings (channels) characterizing areas which create compressive or tensile strain fields. In a similar way, intrinsic crystal interfaces can be formed and stabilized by a two- to three-dimensional network of pentagonal, hexagonal, and heptagonal units. First-order twins contain by far the lowest energy followed by higher-order twins up to the ordinary large-angle grain boundary. In the first two cases (ending with third-order twins) one speaks of coherent grain boundaries if both matching crystal species contain sufficient lattice sites.

Hornstra[212] and Kohn[213] have investigated independently the structure of such interfaces from a predominantly theoretical viewpoint. Whereas Kohn did not allow broken bonds, these were included in Hornstra's work. In Fig. 49 both types of bonding models can be seen[212]. Both authors did not sufficiently point out the "three-dimensional" character of the interfaces studied including the various bond distortions. Mataré[214] has also studied the structural aspects of grain boundaries at different angles. The lower the tilt angle of a certain bicrystal, the more pronounced its "textbook character" is expected in terms of accumulated individual edge dislocations. Investigations on synthesized bicry-

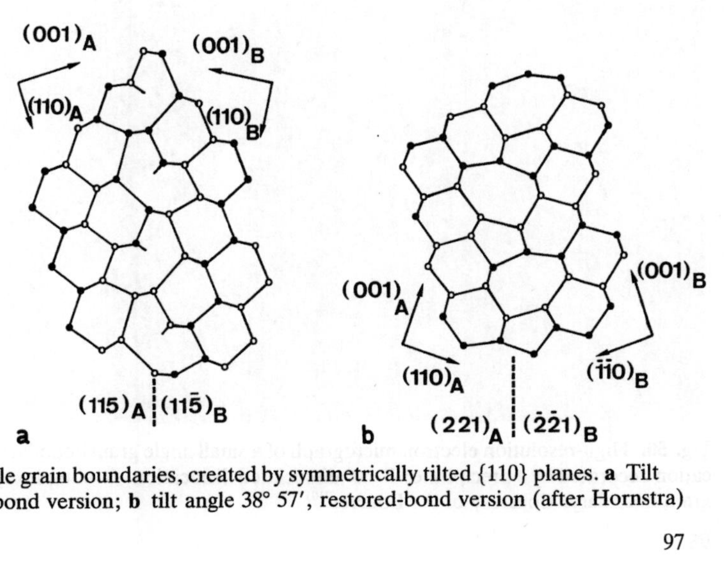

Fig. 49 a, b. Large-angle grain boundaries, created by symmetrically tilted {110} planes. **a** Tilt angle 31° 35′, broken-bond version; **b** tilt angle 38° 57′, restored-bond version (after Hornstra)

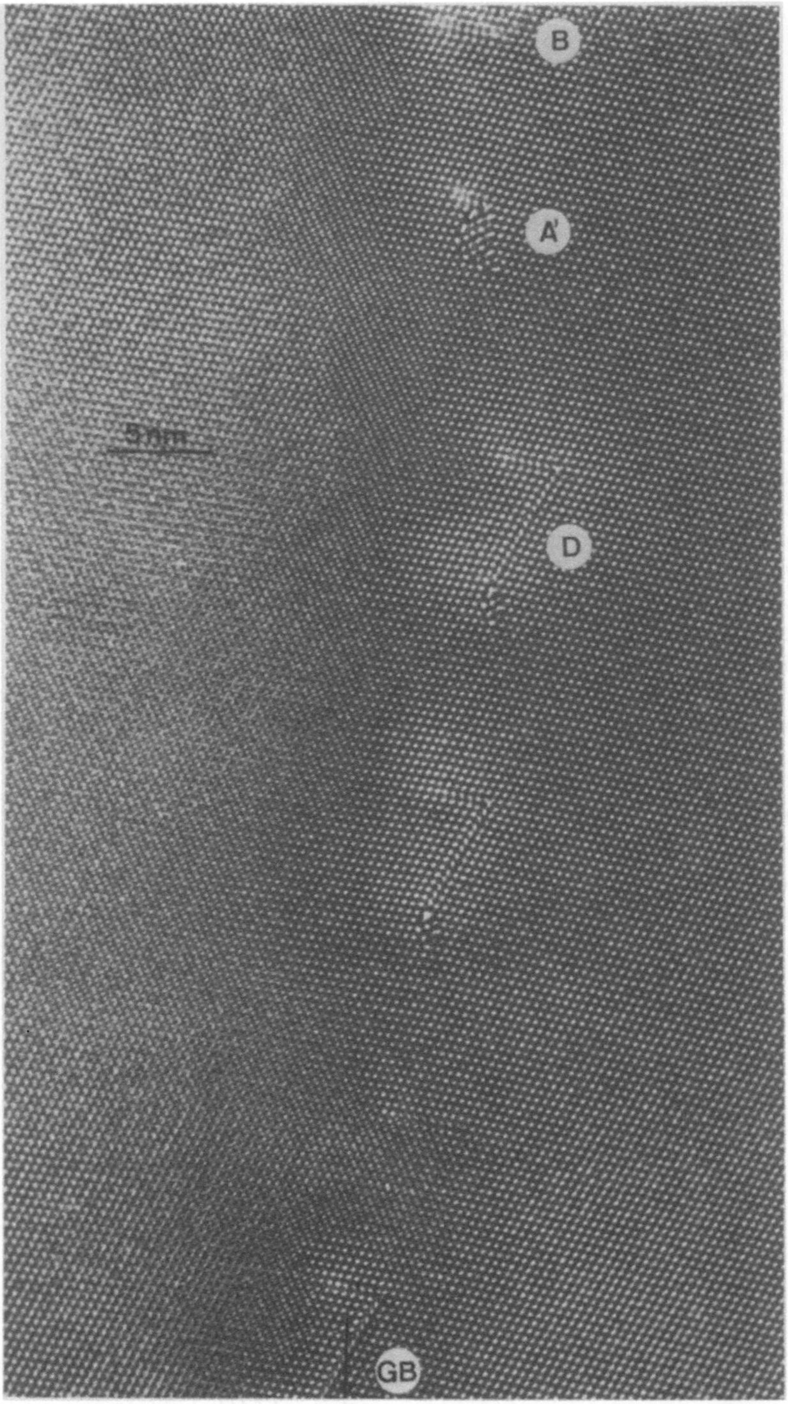

Fig. 50. High-resolution electron micrograph of a small angle grain boundary. A' pure edge dislocation decorated by a precipitate; B 60° dissociated dislocation; D complex type of dislocation; GB grain boundary; Courtesy of A. Bourret[210)]

stals by means of high-resolution electron micrography revealed a rather complex agglomerate of different types of dislocations[210] (Fig. 50). There is no doubt about our present lack of knowledge of practically any type of grain boundary in silicon. The increasing interest in multicrystalline base material for solar cells, however, has started a number of projects which we hope will lead soon to a more profound understanding of this very difficult subject.

6.4.2 Solid-State Reactions

The functionality of electronic devices is affected in various ways by impurities and lattice defects. To eliminate such detrimental effects their mechanisms have to be sufficiently understood, as indicated in Sect. 6.3. In the course of different crystallization processes or during high-temperature treatment in device production a number of solid-state reactions in silicon occur which have not yet been investigated in detail and are thus far from any thorough comprehension. As one of the many examples the formation of depletion zones in the vicinity of high-energy grain boundaries is described here (Fig. 51). This process occurs during annealing (6 h) of heavily contaminated SILSO [C and O both $10^{18}/cm^3$] at 1200 °C[215]. Low-temperature annealing or phosphorus diffusion (at 850 °C) decreases L_{eff} at different rates, depending on crystal perfection and impurity concentration[210].

Attempts to eliminate defect-induced influences on solar-cell performance during device processing have led to different results. In certain cases, the material behaved rather stable through all stages of the pertinent technology while in some others gettering phenomena occurred. At present, three different approaches of this kind, which have been experimentally verified, are being discussed.

The first one, a two-step diffusion process, has been proposed by DiStefano[217]. The first stept involves low-temperature diffusion of phosphorus preferably at grain boundaries. A subsequent high-temperature (drive-in) diffusion forms a highly n-doped skin coating every grain. The resulting p-n junction around each grain interface acts as a

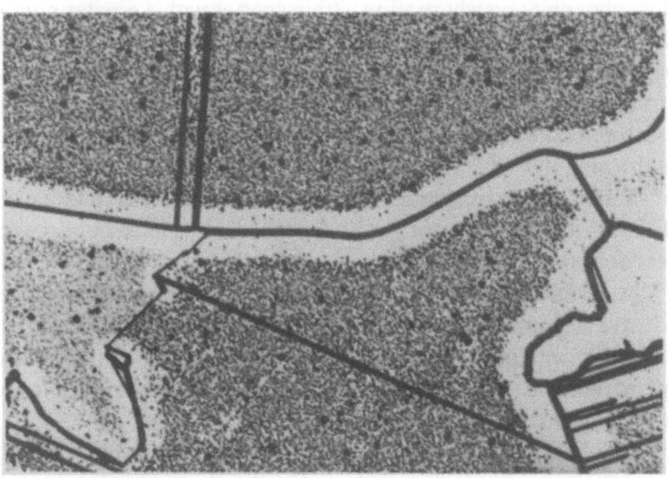

Fig. 51. Heavily contaminated SILSO wafer annealed at 1200 °C/6 h; magnification 90 ×. Secco[233] etch reveals grain boundaries, twin boundaries dislocations, and point-like defects. Depletion of point-like defects in vicinity of high-energy grain boundaries is clearly visible

collector for minority carriers which otherwise might recombine at the grain boundaries. A somewhat modified method has been proposed by Lindmayer[218)]. Further work has emphasized the optimization of diffusion conditions. It has been shown with SILSO material that such two-step processes can reduce the electrical activity of grain boundaries and, consequently, increase conversion efficiency[219, 220)].

The second approach to be mentioned here is largely based on device design. BSF and BSR techniques[221)] have been used to improve CE of high-efficiency cells (HEC BSFR)[222)]. Both variations are generally not applicable to "standard" solar-sheet material due to the necessity of low thickness and/or L_{eff} 100 μm. SILSO with its present 50–100 μm level, therefore, would not appear to be the best candidate here. On the other hand, measurements on "bicrystalline" CZ-grown silicon have indicated that in silicon of even lower L_{eff} values, BSF may be utilized for increasing cell output[223)]. It is possible that BSF, as the special cell-structure type, is not directly responsible for increased PV efficiency but for more alterations in device processing leading to gettering processes which differ from those occurring in standard heat-treatment cycles.

A modified BSF technique is characterized by using the grain boundaries in the p-n region instead in the back surface[224)]. In contrast to the DiStefano process, prediffusion is carried out here with boron instead of phosphorus leading to $n^+/p^+/p$ structures. By these means, L_{eff} in the bulk *and* in the grain boundary areas as well as spectral response data can markedly be improved[225)].

In the third method, the passivation of "dangling bonds" by (atomic) hydrogen is applied to inherent grain boundaries in silicon. The basic principle here has been discussed first for the case of amorphous silicon which is already described in Chap. 5. More recently, it has been proposed to annihilate the electrical activity of as-grown large-angle grain boundaries[226)]. Since this has been achieved experimentally with SOC sheet material such hypothesis seems to be confirmed. After hydrogenation of the devices at 350 °C for 16 hours at 2 torr total pressure, the minority carrier recombination was reduced and the p/n junction quality improved[227)].

It has been pointed out in Sect. 6.4.1 that grain boundaries like dislocations exhibit a dual phenomenon where a low-strain species with a greater number of dangling bonds is the counterpart of a highly strained bond system free of dangling bonds. Under these aspects, a macroscopic grain boundary may simultaneously consist of such alternative structural regions and thus explain the different photoelectric properties of one and the same defect. Passivating experiments with SILSO at higher temperatures using molecular hydrogen (700 °C[228)], 1000 °C[229, 230)]) resulted in improvements of L_{eff}, I_{SC}, and V_{OC} 222 or showed the disappearance of two still identifiable trap levels ($E_C - E_T = 0.115$ V and 0.185 V) in DLTS spectra[229, 230)]. Further studies must be made to elucidate the pertinent reaction mechanisms.

Material alteration by impurity redistribution (e.g. through gettering) is another way to achieve grain boundary passivation. Investigations in this direction have already been started[231)] with good success. Pregettering by the aid of phosphorus diffusion plus subsequent removal of the impurity-enriched glassy layer by etching has proved to be remarkably efficient.

PV CE may also be increased by intermediate gettering[194)]. The good results obtained by such treatment are demonstrated by the photoresponse curves in Fig. 52 where partially refined MG-Si has been used as the base material[194)].

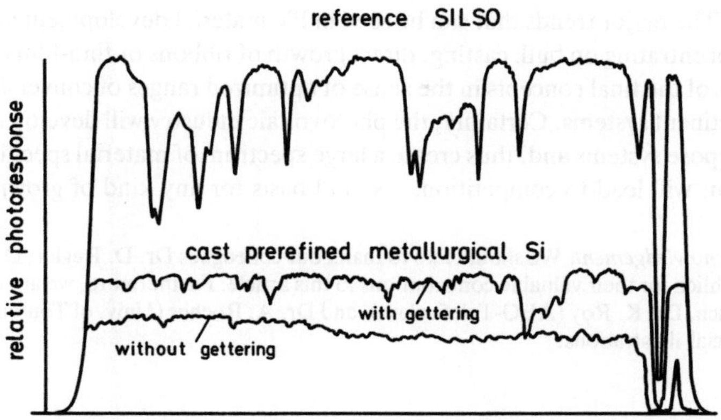

Fig. 52. Relative photoresponse of multicrystalline solar cells from semiconductor-grade SILSO and partially refined metallurgical-grade (PRMG) cast silicon. Standard process cells made from PRMG silicon have shown average efficiencies of 3% AMO and $L_{eff} \leq 5\ \mu m$. High temperature gettering improves efficiency to 5% and $L_{eff} = 10\ \mu m$. The gettered cells show improved photoresponse in the bulk region. Courtesy of AEG-Telefunken[194)]

7 Conclusion

Projects in terrestrial photovoltaics are certainly unique due to the predominant role of their low-cost target. To summarize the needs and trends here, the following criteria have to be considered:
- cost and availability of starting materials
- utilization of by-products
- amount of investment per unit output
- energy consumption.

With regard to by-products, at least three different situations have to be assessed. The first one is to obtain cheap by-prodcuts which may be discharged without major restrictions. Secondly, silicon production will generate valuable by-products of great demand. The third situation is one of increasing importance, namely complete recycling.

Minimizing the amount of investment per annual unit output is one more important economical criterion. The simpler the design of the basic process and the lower the number of necessary processing steps, the better are the chances to reach this goal.

The time necessary to operate the solar cell so as to pay back the energy required to manufacture it is roughly 10 years for terrestrial cells if conventional technologies are used. Improvements in ingot sawing and crystal growth technology and replacement of the Siemens process by a TSG silicon process could reduce the energy pay-back time to at most 12 months including a share of the materials of maximally 6 months [232)]. Certainly, proper choice of heating systems and heating cycles like solid-liquid phase transitions will be of considerable influence. Even when aiming at a silicon substantially lower-priced than 10 $/kg, different routes are prognosticized to be within our reach if we are cost-conscious in every processing step since energy consumption and economics have to be widely considered.

The major trends that can be seen in PV material development at its present stage are concentrating on bulk casting, direct growth of ribbons or thin-film deposition irrespective of the final concepts in the sense of optimized ranges of conversion efficiency in pertinent systems. Certainly, the photovoltaic industry will develop multi-process/multipurpose systems and, thus create a large spectrum of material specifications. This, in turn, will lead to competition, a sound basis for any kind of growing business.

Acknowledgement. We should like to thank our colleagues Dr. D. Regler, Dr. H. Seiter, and R. Wahlich for their valuable contributions to this article. Furthermore, we are indebted to Dr. K-D. Rasch, Dr. K. Roy (AEG-Telefunken), and Dr. A. Rocher (Univ. of Toulouse) for providing special illustrations.

8 References

The following key reference books are marked by symbols in order to facilitate further listing of the vast literature in photovoltaic work

A 11 Proc. 11th IEEE Photovoltaic Specialists Conf., Scottsdale, Arizona (IEEE, New York 1974)
A 12 Proc. 12th IEEE Photovoltaic Specialists Conf., Baton Rouge (IEEE, New York 1976)
A 13 Proc. 13th IEEE Photovoltaic Specialists Conf., Washington, D. C. (IEEE, New York 1978)
A 14 Proc. 14th IEEE Photovoltaic Specialists Conf., San Diego, CA (IEEE, New York 1980)
B 1 Proc. 1st E. C. Photovoltaic Solar Energy Conf., Luxemburg (D. Reidel Publ. Comp., Dordrecht 1978)
B 2 Proc. 2nd E. C. Photovoltaic Solar Energy Conf., Berlin (D. Reidel Publ. Comp., Dordrecht 1979)
B 3 Proc. 3rd E. C. Photovoltaic Solar Energy Conf. Cannes (D. Reidel Publ. Comp., Dordrecht 1981)
C 1 "Semiconductor Silicon 1969", R. R. Haberecht, E. L. Kern (eds.), The Electrochemical Society Inc. Princeton (1969)
C 2 "Semiconductor Silicon 1973", H. R. Huff, R. R. Burgess (eds.), The Electrochemical Society Inc. Princeton (1973)
C 3 "Semiconductor Silicon 1977", H. R. Huff, E. Sirtl (eds.), The Electrochemical Society Inc. Princeton (1977)
D 1 Proc. of Symp. Electronic and Optical Properties of Polycrystalline or Impure Semiconductors and Novel Silicon Growth Methods, K. V. Ravi, and B. O'Mara (eds.), The Electrochemical Society Inc., Pennington (1980)
D 2 Proc. of Symp. Materials and New Processing Technologies for Photovoltaics, J. A. Amick et al. (eds.), The Electrochemical Society Inc., Pennington (1981)

For comprehensive information on photovoltaic materials see
Bachmann, K. J., in: Current Topics in Materials Science, Vol. 3, p. 477, E. Kaldis (ed.), North-Holland, New York 1979

1. Sirtl E., B 1, p. 84
2. Silicon and Ferrosilicon, 2nd edit., Roskill Information Services Ltd., London, 1976 (Anonymous authors)
3. Modified chart taken from Sames, C.-W.: Die Zukunft der Metalle, Suhrkamp Verlag, Frankfurt/Main, 1971
4. Photovoltaic Conversion Program, Summary Report, Energy Research & Development Administration, Washington, 1976 (ERDA-76-161)
5. Hamakawa, Y.: B 3, p. 22 and D 2, p. 18
6. Van Overstraeten, R. J.: D 2, p. 37

7. Photovoltaic Program, Program Summary, January 1978 (DOE/ET-0019/1), Division of Solar Technology, U. S. Department of Energy, Washington, D. C.
8. Wolf, M.: J. Vac. Sci. Technol. *12,* 984 (1975)
9. Sirtl, E.: B 3, p. 236
10. Hopkins, R. H. et al.: DOE/JPL 954 331
11. Dietl, J.: D 2, p. 48
12. Freiesleben, W.: B 3, p. 166
13. Blocher, J. M. jr., Browning, M. F., Wilson, W. J., in: Proc. 7th Int. Conf. on CVD. Sedgwick, T. O., Lydtin, H. (eds.), The Electrochemical Society, Inc., Princeton, 1979, p. 140
14. Matsubara, S. et al.: Proc. 12th Conf. Solid State Devices, Tokyo, 1980, in print
15. McCormick, J. R.: Quart. Rep. DOE/JPL 955533-79/1 (Jan. 1980)
16. Ban, V.: J. Electrochem. Soc. *122,* 1389 (1975)
17. U. S. Patent No. 3, 745, 043
18. Wolf, M.: A 11, p. 306
19. Wolfson, R. G. et al.: Quart. Rep. DOE/JPL 95 4844-78/4 (1978)
20. Rae, S. N.: Final Report DOE/JPL 954 887 (February 1979)
21. Fiegl, G. F.: Quart. Rep. DOE/JPL 95 4886-78/1 (1978)
22. Liaw, H. M.: Symp. Novel Crystallization Techniques, Spring Mtg. Electrochem. Soc., St. Louis, 1980, Ext. Abstr. Nr. 322
23. Kuroda, E., Matsubara, S., Saitoh, T.: Jap. J. Appl. Phys. *19,* L 361 (1980)
24. Fiegl, G. F., Bonora, A. C.: A 14, p. 303
25. Hunt, L. P.: A 11, p. 259
26. Hunt, L. P.: Spring Mtg. Electrochem. Soc., Boston, 1979, Ext. Abstr. No. 59
27. German Patent (DP) No. 11 02 117
28. German Patent (DP) No. 11 40 549
29. Breneman, W. C., Mui, J. Y. P.: Annual Report, DOE/JPL 95 4334-76/3 (October 1976)
30. Breneman, W. C., Farrier, E. G., Morihara, H.: A 13, p. 339
31. Wöhler, F., Liebig Ann. *104,* 94 (1857)
32. Reed, W. H. et al.: A 13, p. 370
33. US Patent No. 4, 150, 248
34. Lyon, D. W., Olson, C. M., Lewis, E. D.: J. Electrochem. Soc. *9 b,* 359 (1949)
35. Blocher jr. et al.: Quart. Rep., DOE/JPL-95 4339-76/2 (October 1976)
36. French Patent No. 2 407 892
37. Sanjurjo, A. et al.: J. Electrochem. Soc. *128,* 179 (1981)
38. Nanis, L. et al.: Final Report DOE/JPL-95 4471 (March 1980)
39. Ingle, W. M., Chaney, R., Rodgers, M.: Quart. Rep. DOE/JPL-95 44 42-77/1 (January 1977)
40. US Patent No. 4, 138, 509
41. Calcote, H. F., Felder, W.: Quart. Rep. DOE/JPL-95 4560-76/2 (December 1976)
42. British Pat. Appl. No. GB 2 028 289 A (Priority Aug. 1978, USA Ser. No. 93 5009)
43. Hunt, L. P. et al.: A 12, p. 125
44. Hunt, L. P., Dosaj, V. D., McCormick, J. R.: Quart. Rep. DOE/JPL-95 4559-76/2 (January 1977)
45. German Patent (DOS) No. 30 13 319 (Priority April 1979, USA, Ser. No. 28 947)
46. Hunt, L. P., Dosaj, V. D.: B 2, p. 98
47. Vigouroux, E.: Compt. Rend. *129,* 334 (1899)
48. US Patent No. 3, 097, 068
49. Brit. Patent Spec. No. 894, 132
50. US Patent No. 4, 124, 410
51. Trumbore, F. A.: Bell System Tech. J. *39,* 205 (1960)
52. Tucker, N. P.: J. Iron and Steel Inst. *15,* 412 (1927)
53. Swiss. Patent No. 567 435
54. US Patent No. 3, 809, 548
55. German Patent (DOS) No. 27 29 464
56. US Patent No. 2, 923, 617
57. US Patent No. 2, 972, 521
58. Hunt, L. P. et al.: Proc. Symp. "Solar Energy", Berkowitz, J. B., Lesk, A., (eds.), Electrochem. Soc., Princeton 1976, p. 200

59. Hurrle, A., Dietl, J.: D 1, p. 106
60. German Patent (DP No. 1, 098. 931
61. US Patent No. 2, 866, 701
62. German Patent (DP) No. 1, 029, 163
63. Dosaj, V. D., Hunt, L. P., Crossman, L. D.: A 11, p. 275
64. Hunt, L. P. et al.: Research and Development of Low-Cost Processes for Integrated Solar Arrays, First Semi-Annual Progress Report (4/15/74-6/30/79), NSF/RANN/SE/GJ-29 729 X/PR/74/2
65. US Patent No. 1, 180, 968
66. German Patent (DP) No. 1, 022, 806
67. German Patent (DOS) No. 2 706 175
68. US Patent No. 4, 200, 621
69. German Patent (DOS) No. 3, 019, 875 (Priority May 24, 1979, USA Ser. No. 41 892, 42 012, and 42 016)
70. Chu, T. L., van der Leeden, G. A., Yoo, H. I.: J. Electrochem. Soc. *125,* 661 (1978)
71. German Patent (DOS) No. 2, 944, 975
72. US Patent No. 4, 193, 974
73. US Patent No. 4, 193, 975
74. US Patent No. 4, 195, 067
75. Runyan, W. R.: Silicon Technology, McGraw-Hill, New York 1965, pp. 33 and references cited there
76. Zerfoss, S. L. et al.: Trans. Faraday Soc. (1949)
77. Schmid, F.: A 12, p. 146
78. Lindmayer, J., A 13, p. 1096
79. Lindmayer, J., B 3, p. 178
80. Ciszek, T. F., Schwuttke, G. H., Yang, K. H.: J. Crystal Growth *46,* 527 (1979)
81. Fally, J., Guenel, C., B 3, p. 598
82. Authier, B., Adv. Solid State Phys. *18,* 1 (1978)
83. Helmreich, D., D 1, p. 184
84. German Patent (DOS), No. 29 25 679
85. Wirth, D. G., Sibold, J. D.: Quarterly Rep., DOE/JPL 954 878-77/1 (Oct. 1977)
86. Kazmerski, L. L., Ireland, P. J., Ciszek, T. F.: Appl. Phys. Lett. *36,* 323 (1980)
87. NL-Pat. No. 74 15 205
88. German Patent (DOS) No. 30 01 815 (Priority Jan. 1979, USA, Ser. No. 4465)
89. Hanoka, J. I., Strock, H. B., Kotval, P. S.: A 13, p. 485
90. Hunt, L. P. et al.: A 13, p. 333
91. Khattak, C. P., Schmidt, F., Hunt, L. P.: D 1, p. 223
92. Schmidt, F., Basaran, M., Khattak, C. P.: ECS Fall Meet. 1980
93. Helmreich, D., Ast, G.: D 1, p. 233
94. US Patent No. 4, 175, 610
95. Chalmers, B.: Principles of Solidification, John Wiley, New York (1964)
96. Fischer, H., Pschunder, W.: A 12, p. 86
97. Yoo, H. I. et al.: A 14, p. 312
98. Roy, K., Pschunder, W.: B 3, p. 263
99. Minnucci, J. A. et al.: A 14, p. 93
100. Green, M. A. et al.: A 14 p. 684
101. Ghosh, A. K., Feng, T., Maruska, H. P.: A 14, p. 1169
102. Ricard, J.: B 1, p. 882
103. Koliwad, K. M., Leipold, M. H.: B 3, p. 228
104. Yoo, H., Schwartz, R. G., Iles, P. A.: A 13, p. 147
105. Fleming, J. R., Holden, S. C., Wolfson, R. G.: Final Rep. DOE/JPL, 954 374 (1979)
106. US Patent No. 4, 187, 827
107. German Patent (DOS) No. 27 22 779
108. German Patent (DOS) No. 27 22 780
109. Anderson, J. R.: A 14, p. 309
110. Pryor, R. A.: A 14, p. 815
111. Lauvray, H., Talpied, A., Besselere, J. P.: B 3, p. 603

112. Khattak, C. P., Schmid, F.: B 2, p. 106
113. Khattak, C. P., Schmid, F.: A 14, p. 484
114. Daud, T., Lin, J. K., Pollock, G. A., Koliwad, K. M.: A 13, p. 142
115. Surek, T.: D 1, p. 173
116. Brissot, J. J., in: Curr. Top. Materials Sci., Vol. 2, (Kaldis, E., Scheel, H. J. (eds.)), North Holland (1977), p. 796
117. Cullen, G., Surek, T. (eds.): Shaped Crystal Growth (Suppl. Issue), J. Cryst. Growth *50* (1980)
118. Gurtler, R. W. et al.: A 13, p. 363
119. Baghdadi, A., Gurtler, R. W.: Ref. 109, p. 236
120. Strunk, H., Ast, D.: Techn. Rep., DOE/JPL 954 852 (April 1980)
121. Yoo, H. I., Iles, P. A.: 4th Internat. Symp. Silicon Materials, Sci. Technol., Symposium Vol.
122. German Patent (DOS) No. 30 17 016
123. German Patent (DOS) No. 26 33 961
124. Kudo, B.: 11th Conf. on Solid State Devices, Tokyo 1979, Tech. Papers, p. 151
125. Kudo, B.: Ref. 109, p. 247
126. Jewett, D. N., Bates, H. E.: A 14, p. 1404
127. Bleil, C. E.: J. Crystal Growth *5,* 99 (1969)
128. US Patent No. 3, 681, 033
129. German Patent (DOS) No. 29 03 061
130. Tsuya, N. et al.: J. Electron. Mater. *9,* 111 (1980)
131. Tsuya, N. et al.: 11th Conf. Solid State Devices, Tokyo 1979, Tech. Papers, p. 153
132. Arai, K. I., Tsuya, N., Takeuchi, T.: A 14, p. 31
133. Ciszek, T. F., Hurd, J. L.: D 1, p. 213
134. Duncan, C. S. et al.: A 14, p. 25
135. Seidensticker, R. G., Hopkins, R. H.: Ref. 109, p. 221
136. Tsivinsky, S. V., Koptev, Y. I., Stepanov, A. V.: Sov. Physics – Solid State *8,* 449 (1966)
137. Duncan, C. S. et al.: DOE/JPL 954 654-79/2 (April 1979)
138. Duncan, C. S. et al.: DOE/JPL 954 654-79/3 (June 1979)
139. Ravi, K. V.: J. Crystal Growth *39,* 1 (1977)
140. Kalejs, J. B. et al.: A 14, p. 13
141. Ciszek, T. F.: Mater. Res. Bull. *7,* 731 (1972)
142. US Patent No. 4, 116, 641
143. Stepanov, A. V. et al.: Bull. Acad. Sci. USSR, Phys. Series *33,* 1826 (1969)
144. Stepanov, A. V.: Soviet Phys. – Tech. Phys. *29,* 339 (1959)
145. Kim, K. M. et al.: Final Rep. DOE/JPL 954 465 (June 1977)
146. Kim, K. M. et al.: Ref. 109, p. 212
147. Kalejs, J. P. et al.: D 1, p. 242
148. Hojo, A., Ikawa, Y., Matsui, T., Nakagawa, M.: Proc. 8th Conf. (1976 Internat.) on Solid State Devices, Tokyo, 1976; Jap. J. Appl. Phys. *16,* Suppl. 16-1, 407 (1977)
149. Kuroda, E., Matsuda, M., Maki, M.: Phys. stat. sol. (a) *48,* 105 (1978)
150. Kuroda, E. et al.: J. Cryst. Growth *43,* 388 (1978)
151. Kimura, K., Koyama, S., Watanabe, H.: A 14, p. 19
152. Duffy, M. T. et al.: Ref. 109, p. 347
153. Maciolek, R. B., Heaps, J. D., Zook, J. D.: J. Electron. Mater. *7,* 441 (1978)
154. Zook, J. D. et al.: Ref. 109, p. 260
155. Belouet, C. et al.: B 1, p. 164
156. Belouet, C.: Ref. 117, p. 279
157. Heaps, J. D. et al.: A 14, p. 39
158. Belouet, C.: D 1, p. 195
159. Chu, T. L. et al.: A 13, p. 1106
160. Robinson, P. H. et al.: A 13, p. 1111
161. Warabisako, T. et al.: Jap. J. Appl. Phys. *17,* Suppl. 17-1, 309 (1978)
162. German Patent (DOS) No. 28 17 797
163. Spear, W. E., Le Comber, P. G.: Solid State Commun. *17,* 1193 (1975)
164. Paul, W. et al.: Solid State Commun. *20,* 969 (1976)
165. Carlson, D. E., Wronski, C. R.: Appl. Phys. Lett. *28,* 679 (1976)

166. Ovshinsky, S. R., Madan, A.: Nature *276,* 482 (1978)
167. Hamakawa, Y., Okamoto, H., Nitta, Y.: A 14, p. 1074
168. Millet, E. J.: J. Cryst. Growth *48,* 666 (1980)
169. Tschöpel, P. et al.: Fresenius Z. Anal. Chem. *302,* 1 (1980)
170. Hill, D. E. et al.: A 12, p. 112
171. Sigmund, H., Branngart, R., Höpfl, Ch.: Verhandl. DPG (IV) *15,* 176 (1980)
172. Holm, C., Sirtl, E.: J. Cryst. Growth *54,* 253 (1981) and to be published in J. Electrochem. Soc.
173. Helmreich, D., Seiter, H.: B 2, p. 742
174. Miller, G. L., Lang, D. V., Kimerling, L. C.: Ann. Rev. Mat. Sci., p. 377 (1977)
175. Weber, J., Wagner, P.: 15th Intern. Conf. Phys. Semicond., Tokyo, 1980
176. Davis jr., J. R. et al.: IEEE Transact. Electr. Dev. *27,* 677 (1980)
177. Salama, A. M.: A 13, p. 496
178. Yoo, H. I., Iles, P. A.: D 1, p. 57
179. Yoo, H. I.: B 3, p. 548
180. Yoo, H. I., Iles, P. A.: ECS Fall Meet. 1981
181. Garrone, L. C. et al., in: Semiconductor characterization techniques, (Barnes, P. A., Rozgonyi, G. A. (eds.)), Electrochemical Soc., Inc., Princeton, 1978, p. 345
182. Yang, K., Schwuttke, G. H., Ciszek, T. F.: J. Cryst. Growth *50,* 301 (1980)
183. Duncan, C. S. et al.: Quart. Rep. DOE/JPL 954 654-79/2 (April 1979)
184. Roy, K.: D 2, p. 1
185. Finetti, M. et al.: Solar Cells *2,* 101 (1980)
186. Soclof, S. I., Iles, P. A.: A 11, p. 56
187. Fischer, H. Pschunder, W.: IEEE Transactions Electron Devices, *Vol. ED-24,* 438 (1977)
188. Ghosh, A. K., Fishman, C., Feng, T.: J. Appl. Phys. *51,* 446 (1980); and literature cited there
189. Ghosh, A. K., Feng, T., Maruska, H. P.: Solar Cells *1,* 421 (1979/80)
190. Roy K., Rasch, K.-D., Fischer, H.: A 14, p. 897
191. Huber, D., Wahlich, R., Helmreich, D.: A 14, p. 316
192. Khattak, C. P., Schmid, F., Hunt, L. P.: D 1, p. 223
193. Schmid, F., Basaran, M., Khattak, C. P.: B 3, p. 252
194. Schmidt, W., Friedrich, G., Rasch, K.-D.: B 3, p. 664
195. Holt, D. B. in: Quantitative Scanning Electron Microscopy (Holt, D. B., Muir, M. D., Grant, P. R., Boswarva, I. M. (eds.)), Academic Press, New York, 1974, p. 213
196. Leamy, H. J., Kimerling, L. C., Ferris, S. D.: Semiconductor Internat., May 1979, p. 75
197. Lile, D. L., Davis, N. M.: Solid State Electronics *80,* 699 (1975)
198. Lile, D. L., Davis, N. M.: SPIE Infrared Technology *62,* 117 (1975)
199. Sopori, B. L., Baghdadi, A.: Solar Cells *1,* 237 (1979/80)
200. Texier, C. et al.: B 3, p. 563
201. Hanoka, J. I.: Solar Cells *1,* 123 (1979/80)
202. Hanoka, J. I., Bathey, B.: A 14, p. 478
203. Schwuttke, G. H., Jang, K., Ciszek, T. F.: J. Cryst. Growth *43,* 329 (1978)
204. Storti, G. et al.: A 14, p. 191
205. Armstrong, R. W. et al.: A 14, p. 196
206. Kazmerski, L. L., Ireland, P. J.: Solar Cells *1,* 178 (1979/80)
207. Kazmerksi, L. L., Ireland, P. J., Ciszek, T. F.: private communication
208. Johan, A. et al.: B 2, p. 750
209. Rocher, A. et al.: A 14, p. 1192
210. Aubert, J. J. et al.: B 3, p. 589
211. Hornstra, J.: J. Phys. Chem. Solids *5,* 129 (1958)
212. Hornstra, J.: Physica *25,* 409 (1959)
213. Kohn, J. A.: Am. Mineral. *43,* 263 (1958)
214. Mataré, H. F.: Defect Electronics in Semiconductors, J. Wiley, New York, 1971
215. Seiter, H.: private communication
216. Legros, B. et al.: B 3, p. 593
217. Di Stefano, T. H., Cuomo, J. J.: Appl. Phys. Lett. *30,* 351 (1977)
218. US Patent No. 4, 230, 508
219. Ferraris, F. et al.: B 3, p. 625

220. Huber, D., Wahlich, R.: private communication
221. Gay, C. F.: A 13, p. 444
222. Rasch, K.-D. et al.: A 14, p. 141
223. Arienzo, M., Muleo, A., Orr, W. A.: B 3, p. 650
224. Orr, W. A., Califano, F. P.: A 14, p. 1198
225. Orr, W. A., Mabilia, G., Paraggio, V.: B 3, p. 646
226. Seager, C. H., Ginley, D. S.: Appl. Phys. Lett. *34,* 337 (1979)
227. Seager, C. H., Ginley, D. S., Zook, J. D.: Appl. Phys. Lett. *36,* 831 (1980)
228. del Valle, J. L., Flores, C., Dueñas.: A 14, p. 202
229. Qualid, J., Amzil, H., Lauer, A.: ECS Spring Meet., St. Louis, May 1980
230. Qualid, J. et al.: B 3, p. 938
231. Chitre, S. R., Donon, J.: B 3, p. 608
232. Hunt, L. P.: A 12, p. 347
233. Secco d'Avragona, F.: J. Electrochem. Soc. *119*, 948 (1972)

The Capillary Action Shaping Technique and Its Applications

Ted F. Ciszek

Solar Energy Research Institute Golden, Colorado 80401, USA

Crystals: Growth, Properties and Applications 5
© Springer-Verlag Berlin Heidelberg 1981

A. Introduction

Conventional melt-growth techniques such as float-zoning, Czochralski pulling, or Bridgman growth typically produce crystals in the form of cylindrical ingots. For many applications, only a thin segment of the crystal is required and hence, the ingot is sectioned into wafers. This operation is usually carried out with a diamondedged saw blade by abrasive cutting action. For small-scale usage or for inexpensive crystal materials, this is a generally acceptable processing step. One undesireable consequence is the introduction of surface damage due to the abrasive contact. However, the damage can usually be removed by appropriate etching and polishing procedures.

For large-scale use of expensive crystal materials, the sawing process is more objectionable. It is difficult to saw thin wafers from an ingot without wasting at least half of the material in the form of fine powder which is ground away as the diamond blade passes through the crystal. Furthermore, the sawing process is time consuming and adds additional costs to processing the material. For these reasons, and to attain other crystal shapes not readily produced by sawing or cutting operations, a technology has evolved for the growth of shaped crystals – especially sheets or ribbons – from the melt.

Early experiments on shaped growth of metal crystals (Cd, Zn, Al, Sn and Bi) were conducted by von Gompertz[30)]. The meniscus shaping was provided by holes in mica plates which floated on the melt surface. Stepanov[21)] elaborated upon this technique, producing crystals of many additional materials and diverse shapes. Numerous melt containment methods, melt level control techniques, shaper materials, and shaper designs were investigated. A summary of that work is given in Stepanov[22)].

One particular area of shaped crystal growth which could have large scale applicability is the growth of semiconductor sheets or ribbons. Thus, several attempts have been made to apply Stepanov's technique to germanium and silicon. For example, germanium ribbons were grown by Brissot and Raynaud[6)], Tsivinskii and Stepanov[27)], and Swets[25)]. Brissot and Raynaud studied orientation effects in 1.5 mm thick ribbons up to 30 mm wide. The later workers found that stable thinner ribbons (0.2 mm) could be grown provided that the ribbon edges were allowed to be thick (0.8–2 mm). Swets achieved growth rates of up to 20 mm/min. In all cases, a non-wetting graphite shaper was used. Boatman and Goundry[5)] grew silicon ribbons with nonwetting shapers (SiO_2, BN, and others) from an RF-heated melt atop a solid silicon pedestal. They were able to produce dislocation-free ribbons up to 12 mm in width but experienced difficulty in growing larger widths. In general, the Stepanov technique involves meniscus shaping through a non-wetting die or shaper, with solidification of the crystal occurring a short distance from the shaper.

Another technique for ribbon growth of some materials is dendritic solidification. Again, interest has mainly been in semiconductor crystals such as germanium and silicon. Billig[3)] introduced controlled dendritic growth of germanium, Bennett and Longini[2)] refined the technique to allow rapid (150 mm/min) growth of 0.25 mm thick by 3 mm wide strips from a supercooled melt. Such growth can only be conducted with materials containing a twin plane parallel to the low surface-free-energy surface which is (111) for Ge and Si. The presence of the twin plane causes an anisotropy in the ease of growth for various directions in the plane, with the <$\bar{1}\bar{1}2$> type directions being the easy growth directions. Thus dendritic Ge and Si ribbons always have a (111) surface, a <$\bar{1}\bar{1}2$> axis, and at least one internal twin plane parallel to the surface. A novel extension of dendritic growth is dendritic web ribbon growth (see Dermatis and Faust[12] or Barrett et al.[1]). Here, two-co-planar dendrites containing the same (111) twin plane are propagated in a supercooled melt. The dendrites are pulled upward from the melt at a rate equal to their growth rate, and a liquid meniscus solidifies into a continuous web between the two dendrites. Ribbons of silicon 30 mm in width and 0.1 mm thick were produced by this technique with a flatness of less than 2.5 μm. In a variation of this method by Tucker and Schwuttke[28], it was possible to grow dislocation-free webs by causing the twin plane to lie within the dendrite edges but outside of the web proper.

Recently Ciszek and Hurd[9a] and Ciszek et al.[10a] described an edge-supported pulling (ESP) sheet growth method for silicon in which two parallel filaments of graphite or quartz are moved upward through liquid silicon while a sheet of silicon solidifies between them, as in dendritic web growth. Unlike web growth, however, a precise thermal environment for controlled dendrite proparation is not necessary. Very stable growth was observed for ribbons 50 mm wide and 0.07 to 2.0 mm thick.

Bleil[4] introduced a horizontal ribbon growth technique for Ge and ice which is currently being applied to silicon. In this method, the ribbon is pulled horizontally from the top surface of the solidifying liquid in a brimful crucible. Provisions are made to maintain the brimful condition and to achieve local cooling in the solidification region. The principal attraction of this method is that the solid/liquid interface extends over a large area even though the ribbon thickness is small. Thus, the heat of fusion is much more readily removed and linear growth rates can be high. Silicon growth rates of 300 mm/min have been reported by Koyanagi[14]. The major drawback of the process is control difficulty.

Neither dendritic ribbon growth nor horizontal ribbon growth involve the use of shaping dies. Hence, a high purity can be maintained in the large melts from which solidification occurs and reasonably good crystal perfection has been achieved. There is another shaper-free ribbon growth process which has been developed by Gurtler et al.[13] for silicon. It involves the use of scanned laser heating to produce a narrow molten zone for ribbon to ribbon float-zoning. To date, this technique has not achieved highly perfect crystals, but high growth rates (125 mm/min) and 50 mm widths have been obtained. Unlike Stepanov's shaper method which produces a variety of crystal cross sections, the last three techniques which were discussed are primarily limited to sheet or ribbon growth.

Another shaper-technique which involves a wettable shaping die was introduced by Tsivinskii et al.[26]. Here, the liquid rises by capillary action in vertical grooves until it reaches the die tip which is some distance above the melt level in the crucible. Tsivinskii used slotted tungsten dies to stabilize the meniscus edges for Ge ribbon growth, while the middle of the ribbon solidified as a web much like the dendritic web growth process mentioned earlier. The ribbon edges which grew from a narrow meniscus defined by the die top edges were 1.25–3.5 mm thick and the middle of the ribbon was 0.1–3.0 mm thick. Ribbon widths of 15–45 mm were achieved at growth rates of 3 to 5 mm/min, and single crystals were reported. LaBelle[15] ex-

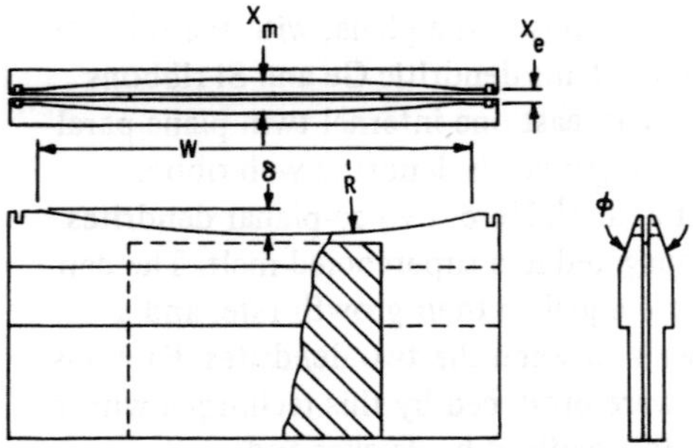

Fig. 1. Capillary Action Shaping Technique (CAST) die design

tended Tsivinskii's wettable shaper method to the growth of diverse cross sectional shapes of sapphire from molybdenum dies. In his work, the shaping dies were fabricated in such a way that all of the growth front was over a narrow meniscus at the die top (rather than just the edges as in Tsivinskii's work). Again, grooves or channels were used to supply new liquid to the meniscus at the die-top via capillary action. Since all growth was from a thin meniscus, the base of which was anchored at the outer edges of the shaped die top, the grown cross-sectional shape closely followed the die top shape. LaBelle termed this process EFG (edge-defined, film-fed growth). In 1972, the EFG method was applied to silicon[7] and most recent work on this method has been on silicon for photovoltaic applications. Reviews with numerous literature references have been made by Ravi[17] and Surek et al.[24]. Other materials currently being studied include the original sapphire and also various metal eutectics.

The capillary action shaping technique which will be discussed in this article contains elements of both LaBelle's[15] approach and Tsivinskii's[26]. Capillary action is used to move new liquid to the top of the shaping die. However, the die-top shape is not closely coupled to the cross-sectional shape of the growing crystal. Substantial improvements in ribbon surface quality, for example, can be achieved by using a higher central melt meniscus than that attainable with the EFG technique. To achieve the higher meniscus, a ribbon die design of the type shown in Fig. 1 can be

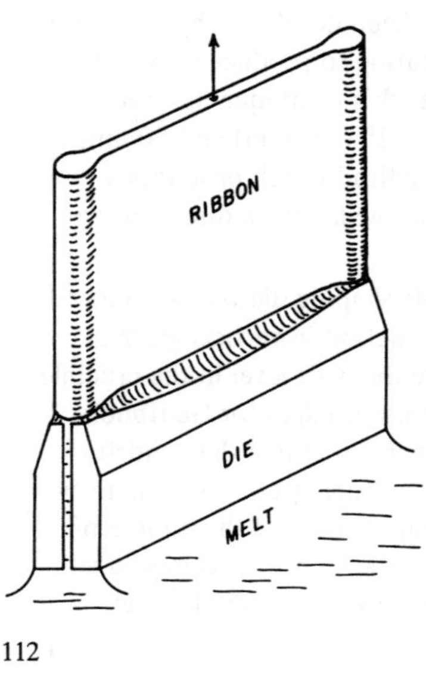

Fig. 2. CAST high melt meniscus growth

used. The ends of the die are closely coupled to the ribbon via a narrow meniscus as in both Tsivinskii's approach and EFG. But the meniscus in the mid-region is much higher (Fig. 2). In the mid-region, the edges of the die are not parallel or concentric with the sides of the growing ribbon. Rather, they form a broad base for the meniscus and allow a greater meniscus height. This results in a smoother ribbon surface because the solid/liquid interface is farther from disturbing influences at the die top such as irregular edges, particulate deposits and mechanical disturbances. In the case of tube or cylinder growth, a high meniscus and loose coupling between die-top and solid liquid interface can be attained with a flat-top die by enlarging the die-top annulus and/or diameter which again provides a broad base for the high meniscus.

In this article, the capillary action shaping technique will be described with particular reference to the growth of silicon ribbons and tubes. Subjects such as growth rates, die design, apparatus, growth techniques, characterization, and material properties and problems will be discussed.

B. Apparatus

Experimental work can be conducted using apparatus of the type described by Ciszek and Schwuttke[10)]. This equipment employed a hydraulically actuated pulling mechanism with a 1.37 m maximum stroke. A pressureized air-over-oil reservoir system provided the driving force for actuation of the moving platen upon which the seed holding shaft was attached. X-Y-Z adjustment was provided for the seed shaft. Platen speed adjustment was attained via a needle valve in the oil flow line between the hydraulic cylinder and the oil reservoir tank. The crystal growth furnace and associated vacuum system were supported on the frame of the hydraulic puller as shown in Fig. 3. Power was supplied by a 50-kW, 450 kHz-rf generator. The furnace consisted of a single-walled 94-mm-diameter silica shell with water-cooled stainless-steel end fittings sealed by viton O-rings. The top fitting could be either completely sealed (for vacuum bake-out of dies and furnace components) or provided with a pull port through which the growing crystal passed. The bottom fitting connected the shell to a horizontal base on one arm of a "tee" fitting. A horizontal sliding plate on an X-Y-θ stage was mounted on the opposite arm of the "tee". This plate held the pedestal shaft upon which the crucible assembly rested. The plate also supported an optical temperature sensor which was focused on the bottom of the crucible holder and was part of the closed-loop temperature control system. The third arm of the "tee" was used for evacuation and as a purge gas inlet.

The furnace hot zone is depicted in Fig. 4. All interior components are graphite except the opaque quartz insulator and the clear quartz crucible. The graphite die and hot-zone components were baked out at 1500 °C for one hour in vacuum prior to use. After bake-out, the quartz crucible was loaded with silicon. Doping was achieved with a dopant/silicon alloy powder which was put into the crucible along with the silicon charge. Growth was conducted in an argon ambient. After the silicon charge had melted, growth was initiated by touching a silicon seed to the melt at the die top. A short portion of the seed was melted back, and then the pulling speed was gradually increased and the temperature adjusted to first neck the growth to a width less than that of the seed and then widen it to the size dictated by the die-top geo-

Fig. 3. Growth apparatus

metry. For a die of the type shown in Fig. 1, the width is limited by the small notches near each end of the die-top. These provide a geometrical boundary without a corresponding abrupt thermal discontinuity.

Since a tube can be considered to be an "edge-less" ribbon, its growth allows a study of the relationship between growth rate and thickness which is independent of width. The growth procedure for tubes was similar to that for ribbons.

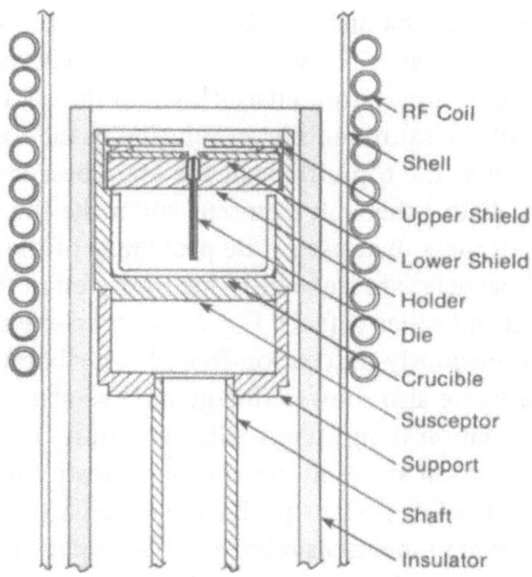

Fig. 4. Furnace hot-zone components

After the ribbon or tube reached the desired length, growth was terminated by increasing temperature and pulling speed to separate the shaped crystal from the liquid meniscus at the die-top. By reseeding, several pulls could be made from the same charge. At the end of the day, the die was salvaged by freezing a seed crystal to the die top and using it to pull the die assembly (die, holder, shields) free from the liquid in the crucible. Although the frozen silicon within the die frequently cracked upon cooling, the die itself was typically crack-free.

C. Capillary Dies

I. Die Material Considerations

The use of wettable shaping dies imposes diverse idealized characteristics upon the die material properties. A low contact angle between the melt and the die is desired, but at the same time the die should be insoluable, non-reactive and physically durable. In practice, this combination is almost impossible to fully achieve. LaBelle[15)] used molybdenum dies for sapphire growth with good results. Ciszek[7)] introduced high density graphite and silicon carbide as die materials for silicon growth. To date these are still the most promising materials. However, the solubility of C in molten silicon is approximately 100 ppma near the melting point according to Scace and Slack[18)]. The effective segregation coefficient of C in silicon is reported by Tucker[29)] to be about 0.05 for normal freezing. Thus carbon impurity levels on the order of 5–100 ppm are expected in silicon ribbons grown from carbon containing dies. Thin long capillary channels and fast growth rates tend to make the effective segregation coefficient close to unity for wettable die growth.

Carbon supersaturation becomes especially high in the melt at the die-top both because solute rejected at the solid/liquid interface cannot readily diffuse down the narrow die channels and because the die-top is the coolest region in the melt. Thus, melt in the crucible which is saturated with carbon from contact with the die-bottom becomes supersaturated as it rises tp the cool die-top. Here, excess carbon is forced out of the saturated silicon solution in the form of β-SiC crystallites, which tend to collect at the top surface of the die. These crystallites, up to a few tenths of a mm in size, distort the melt meniscus and make the ribbon non-uniform in its surface smoothness. Because of the proximity of the freezing interface to the die top, the SiC particles are frequently incorporated in the ribbon, where they generate dislocations and other defects. Fig. 5 shows an example of a burst of dislocations generated in a previously dislocation-free silicon ribbon by the inclusion of a large SiC particle. The figure also shows subsequent large-area stacking fault nucleation, twinning, and degradation to multicrystalline structure. It is conceivable that the lattice mismatch near carbon atoms in the crystal nucleate these latter defects. As will be discussed later, the SiC particle problem can be partially alleviated by die designs which achieve a high melt meniscus relative to the typical particle diameter. However, it would be desireable to have a die material which is less reactive with silicon.

Several potentially useful materials have been tested under actual ribbon growth conditions. Hot pressed "high purity" AlB_{12}, TiB_2, ZrB_2, Al_4C_3, B_4C, and Si_3N_4

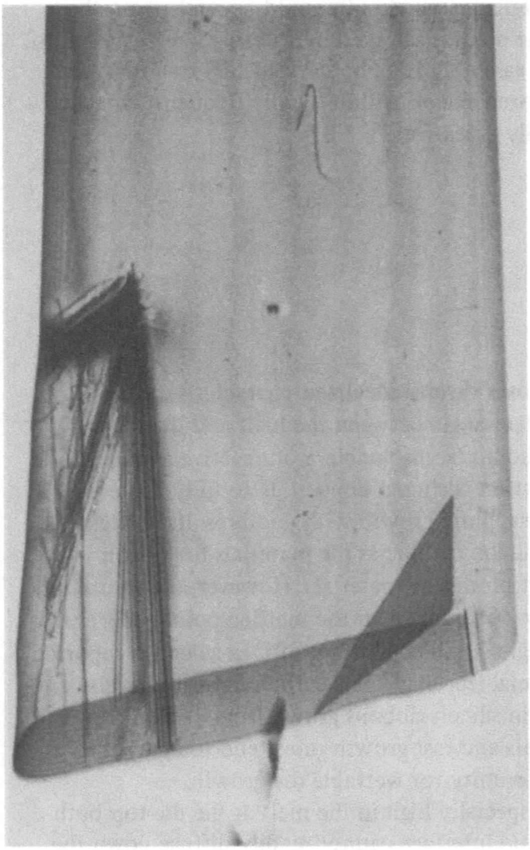

Fig. 5. X-ray topograph showing the nucleation of a dislocation burst in a previously dislocation-free silicon ribbon by a SiC particle. Further degradation by large-area stacking faults, twinning and random grain formation is also depicted

Table 1. Manufacturer's data for vitreous carbon

Chemical characteristics	
Maximum operating temperature	> 2,500 °C
Apparent density	1.50 to 1.55 g/cm^3
Apparent porosity (no absorption of oil and mercury)	0
Permeability to gas; at normal temperature	better than 10^{-8} cm^2/s
at 2,500 °C	approx. 10^{-6} cm^2/s
Ash content	0.005 % (50 ppm)
Sulphur content	< 50 ppm
Boron content	< 2 ppm
Oxidation resistance at 600 °C	0.1 mg/cm^2/h
at 800 °C	2 mg/cm^2/h
Mechanical properties	
Transverse breaking strength	600 to 800 mg/cm^2
Compressive strength	1,500–2,000 kg/cm^2
Shore hardness	120–125
Hardness Moh's scale	6–7
Vickers micro-hardness (30 kgm)	150–175
Young's modulus	2,200 kg/mm^2
Thermal properties	
Coefficient of thermal expansion: at 100 °C	3.2 x 10^{-6} per °C
at 650 °C	3.5 x 10^{-6} per °C
Thermal Conductivity	
Resistance to thermal shock	0.06 cal/cm/s/°C
Air quenching acceptable from ... to room temperature	2,500 °C
Electrical properties	
Electrical Resistivity	4,500 microhm/cm

were obtained in the form of rectangular blocks 3 x 10 x 30 mm in size. No binder was used in hot pressing. Purity levels were 99–99.9%, densities were about 90% of theoretical densities and the starting powder particle size was typically 2–5 μm, according to the manufacturer. Vitreous carbon was obtained in the form of 2 mm thick plates, with properties as shown in Table 1. Two sources of silicon carbide were used. One was Crystar silicon carbide from Norton Industrial Ceramics Division, the other was KT silicon carbide from the Carborundum Company. A silicon nitride-bonded silicon carbide was also obtained from Carborundum. The manufacturers data for these three materials is given in Table 2. The Crystar material is nominally 99% pure recrystallized silicon carbide, with Ti, Cr, Co, V, and Al as typical metallic impurities. Purity levels of the carborundum materials are not known at this time. Pyrolytic boron nitride was also investigated.

II. Fabrication and Evaluation of Experimental Die Materials

Dies were fabricated from all materials except aluminium carbide, which had crumbled during storage, and boron nitride. The dies were made 9.5 mm wide with a

Table 2. Manufacturer's data for various silicon carbides

Property	KT silicon carbide	REFRAX silicon nitride-bonded silicon carbide	CRYSTAR silicon carbide
Apparent specific gravity (g/cm^3)	3.10	2.65	2.6
Working temperature			
Inert atmosphere	2310 °C max.	1760 °C max.	—
Oxidizing atmosphere	1650 °C max.	—	1760 °C max.
Modulus of rupture (Pa)	1.57×10^8 (1200 °C)	$3.5–4.8 \times 10^7$ (1370 °C)	$7–15 \times 10^7$ (1500 °C)
Modulus of elasticity (Pa)	3.9×10^{11} (21 °C) 3.5×10^{11} (1200 °C)	1.2×10^{11} (21 °C)	2.0×10^{11}
Compressive strength (Pa)	1.0×10^9 (21 °C)	1.4×10^8 (21 °C)	6.9×10^8
Tensile strength (Pa)	1.42×10^8 (21 °C)	$2.1–2.4 \times 10^7$ (21 °C)	—
Thermal conductivity (W/m°K)	104 (615 °C mean temp) 67 (857 °C mean temp) 36 (1141 °C mean temp) 25 (1256 °C mean temp) 14 (1528 °C mean temp)	16.4 (1204 °C)	26 (538 °C) 24 (816 °C) 21 (1260 °C) 19 (1593 °C)
Thermal coefficient of expansion (cm/cm°C x 10^{-6})	3.38 (20– 430 °C) 4.68 (20– 980 °C) 5.04 (20–1370 °C)	3.2 (20– 200 °C) 4.7 (20–1370 °C)	4.9 (20–1425 °C)
Mean specific heat	0.31 (20–1540 °C)	0.29 (20–1370 °C)	0.35 (1650 °C)
Spectral emissivity at 870 °C	0.9	0.9	—
Permeability	Impermeable	Negligible	—
Porosity (%)	0	15	18

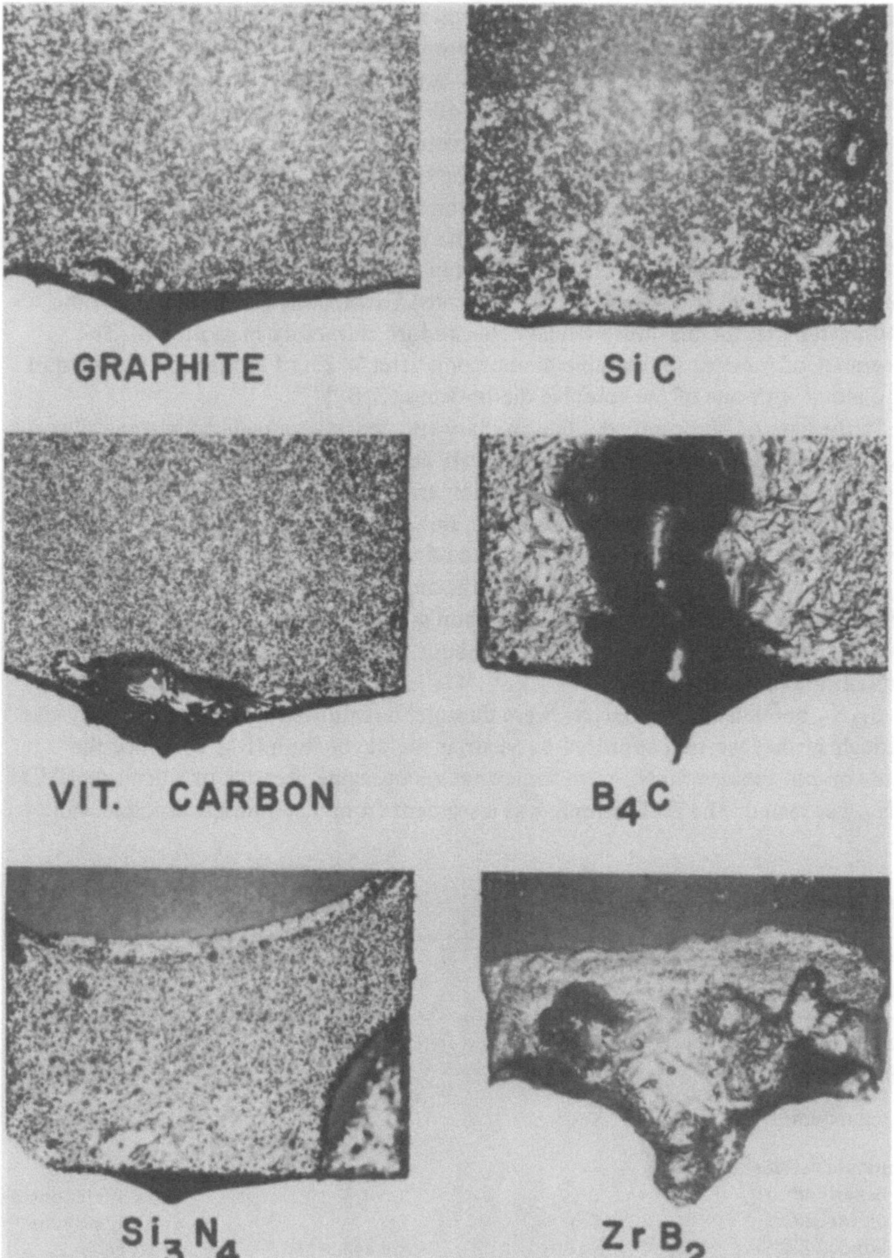

Fig. 6. Die bottoms after contact with molten silicon for more than two hours

0.76 mm top thickness and a 0.56 mm capillary slot thickness. All machining was done with 180 grit diamond wheels. No serious machining problems were encountered, except that vitreous carbon and silicon nitride-bonded silicon carbide chipped easily. Slow machining rates were necessary on silicon carbide and boron carbide to avoid chipping. After machining, the dies were baked out in vacuum at 1500–1600 °C for one hour. The AlB_{12} die outgassed severely during bakeout and the temperature could not be raised above 1400 °C without causing the pressure to rise above 2×10^{-4} Torr. This die was considerably erroded and broke after bake-out. The Si_3N_4 die was a light tan color – the others were grayish to black.

The growth procedure described in section B was used in attempting ribbon growth with the experimental materials. If the die did not fill with liquid silicon, the system temperature was increased considerably above normal levels to attempt to encourage melt rise. If melt rise occurred, crystal growth was attempted. After the growth attempts, the dies were visually checked for durability in molten Si. The appearance of some of the durable die-bottoms after $\geqslant 2$ h of immersion is shown in Fig. 6 along with one of the soluable die-bottoms (ZrB_2).

In the case of boron nitride, the capillary rise check was made by placing silicon in a boron nitride crucible with a slotted wall, and melting the silicon.

The results of the die material evaluation are summarized in Table 3. For the particular forms of the die materials used here, only boron carbide, KT silicon carbide, Crystar silicon carbide, vitreous carbon, and of course, high density graphite were adequately wettable and durable to allow the growth of long ribbons (> 0.5 m). In addition, TiB_2 and ZrB_2 allowed short ribbon growth, but the dies suffered acute dissolution as can be seen in Fig. 6. The ribbons grown from experimental die materials are pictured in Fig. 7.

Si_3N_4 bonded SiC, and Si_3N_4 were durable, but showed no sign of capillary rise although surface wetting appeared to occur at the die bottom (Fig. 6). Since the results on hot-pressed Si_3N_4 were somewhat encouraging, wetting of silicon on CVD Si_3N_4 was tested. The CVD sample was a segment from a cylindrical crucible wall.

Table 3. Die material evaluation

Material	Capillary rise	Durable in molten Si	Ribbon growth
Aluminum boride[a)]	—	—	—
Titanium boride	Yes	No (dissolves)	Short ribbons
Zirconium boride	Yes	No (dissolves)	Short ribbons
Aluminum carbide[b)]	—	—	—
Boron carbide	Yes	Yes	Long ribbons
Silicon carbide	Yes	Yes	Long ribbons
SiN bonded SiC	No	somewhat	None
Vitreous carbon	Yes	Yes	Long ribbons
Graphite (dense)	Yes	Yes	Long ribbons
Boron nitride	No	Yes	None
Silicon nitride	No[c)]	Yes	None

a) Vapor pressure too high at 1412 °C
b) Decomposed and crumbled during storage
c) Evidence of surface wetting, but no rise in capillary slot

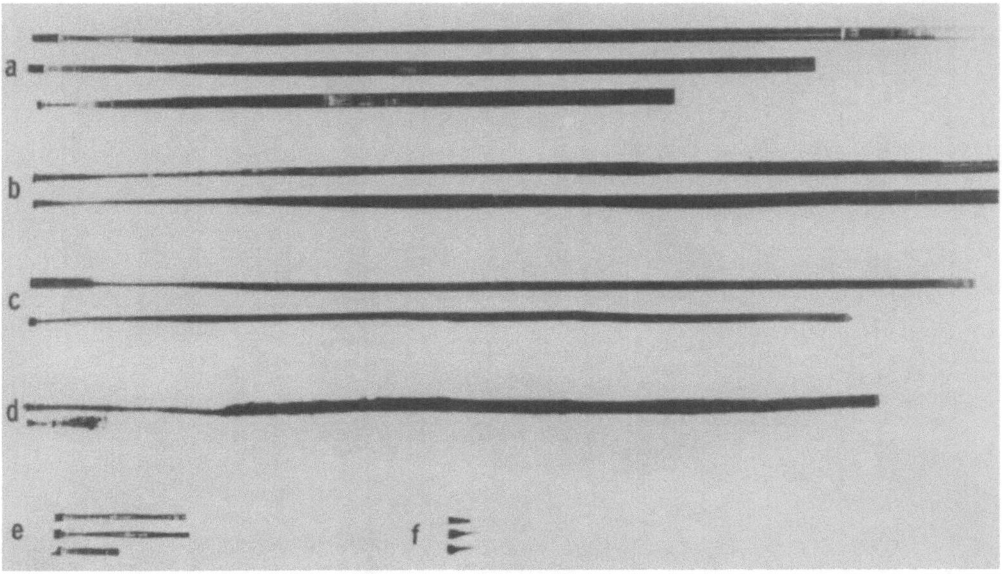

Fig. 7 a–f. Silicon ribbons grown from experimental die materials: (a) KT SiC, (b) Crystar SiC, (c) Vitreous carbon, (d) B_4C, (e) TiB_2, (f) ZrB_2

The radius of curvature (in only one dimension) was 4 cm, and the segment thickness was 1.1 mm. The Si_3N_4 segment was degreased, etched in HF for 2 min and rinsed. An etched silicon chip of 0.078 g weight was placed on the Si_3N_4 and heated to the silicon melting point under an argon ambient. The droplet was allowed to remain molten for several minutes and then was solidified. The droplet and substrate were then sectioned and etched to reveal the interface and contact angle; a close-up of one edge appears in Fig. 8. The observed contact angle is 23 °. The partial wetting behavior indicates that capillary rise should occur with CVD Si_3N_4. While the main body of the droplet had the contact angle described above, a thin film of silicon spread on the Si_3N_4 ahead of the main body.

Pyrolitic boron nitride showed no sign of wetting or capillary rise, and it was not appreciably attacked by liquid silicon.

The B_4C die produced strong surface morphology features which are probably a result of excessive boron pick-up. The resistivity of ribbons grown from this die was approximately 0.004 Ω-cm. The ribbon surface was very clean, however. Wetting and capillary rise were excellent with this die.

Vitreous carbon dies were similar to graphite dies in their performance. Wetting and capillary rise were very good, the die-bottom showed good resistance to dissolution, and the melt in contact with the die remained very clean. Vitreous carbon is also a promising material for use as a crucible both for ribbon growth and for directional solidification (see Ciszek et al.[11]). As with graphite dies, SiC particles formed in the meniscus near the top of the vitreous carbon dies and were incorporated in the surface of the growing ribbon.

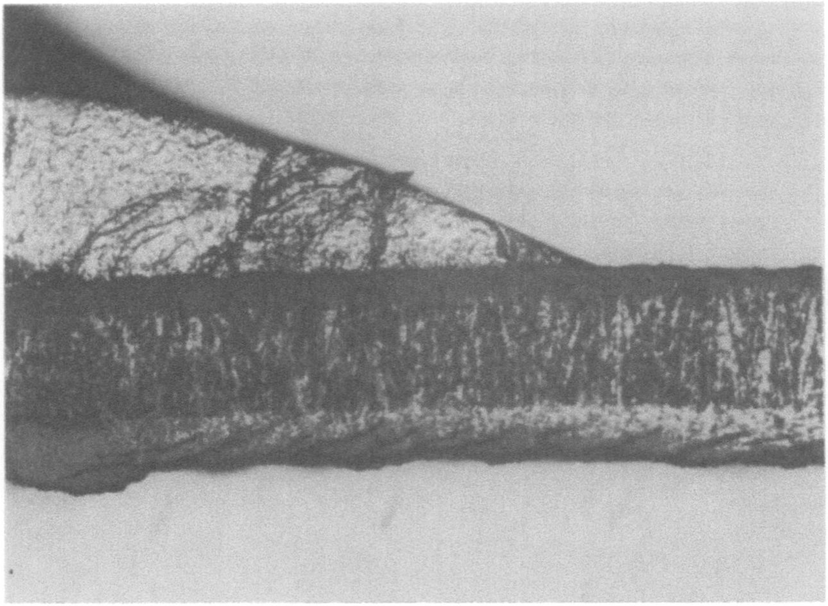

Fig. 8. Close-up (25X) of one edge of a frozen Si droplet on a CVD Si_3N_4 substrate

The ribbons grown from silicon carbide dies were quite free of surface particles in the early stages of growth and some large single crystal regions were seen. Later in growth, these ribbons also incorporated surface particles, but at a low density. A distinct difference in the morphology of silicon carbide particle formation at the die top was seen between graphite and silicon carbide dies. In the case of graphite dies, individual yellow β-SiC crystals form at the top and sides. With the KT silicon carbide dies, a dense, continuous film of SiC tends to grow from the carbon-saturated silicon melt near the die top. When the die is first used, this film does not interface appreciably with the Si ribbon growth. The initial ribbon perfection (both with KT SiC and with Crystar SiC) can be quite good. As ribbon growth progresses, however, the SiC film also grows and, at some stage begins to break loose from the die top in clumps, which are then incorporated in the ribbon and disturb its perfection. The frequency of this occurrence is similar to that of SiC crystallite incorporation with graphite dies, and the appearance of the grown ribbons is not too different from that of ribbons grown with graphite dies.

The different type of SiC formation at the die-top seen with SiC dies may be due to the nature of the nucleation sites for crystallization of excess SiC from the carbon-saturated liquid silicon at the SiC die top. If the SiC die material promotes better epitaxial adherence of SiC grown from solution than does graphite, then less free particulate SiC would be available in the meniscus region for incorporation in the growing ribbon. If this concept is valid, better results would be expected from pure, dense SiC dies. A 50 mm wide die of two-piece design (Fig. 9) was made of graphite and an impervious CVD SiC coating was applied. The die was tested for long term performance. A single die has been used for 24 growth attempts in 14 melt-down

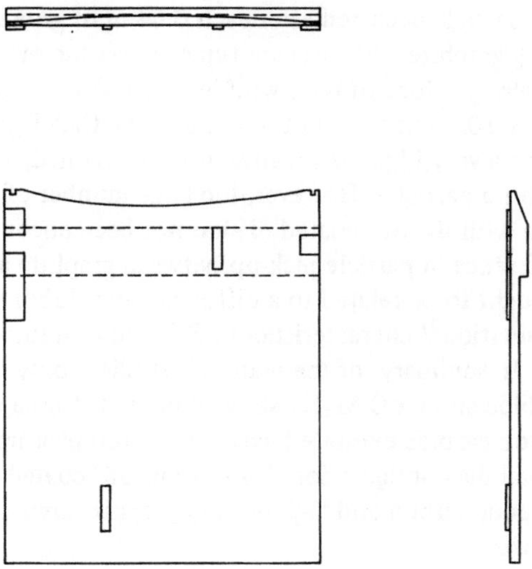

Fig. 9. Two-piece capillary die design

cycles and has produced 17 ribbons of total length 13.1 m. The SiC particle density for individual consecutive ribbons and also the accumulative density (total SiC particles/total area grown) are summarized in Fig. 10. The accumulative SiC particle density was $1.3 \times 10^{-3}/cm^2$ after 2×10^3 cm^2 of ribbon had been pulled and rose to $9 \times 10^{-3}/cm^2$ after 4.5×10^3 cm^2 of ribbon had been grown from the die. After 6.5×10^3 cm^2 of growth, the accumulative particle density was $1.6 \times 10^{-2}/cm^2$. The rise in accumulative particle density is largely due to two ribbons which had abnormally high densities of $7.1 \times 10^{-2}/cm^2$ and $1.3 \times 10^{-1}/cm^2$. One of these occurred during a run in which the graphite heat shields were replaced.

Discounting the two high density ribbons, the accumulative SiC particle density was $3.7 \times 10^{-3}/cm^2$. Another way of expressing this result is that, on the average,

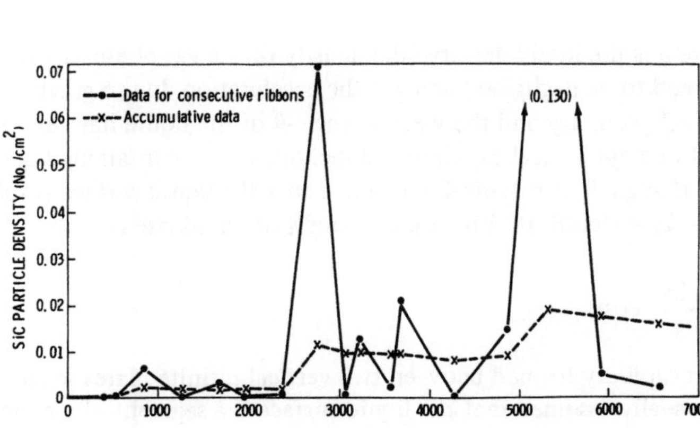

Fig. 10. Silicon carbide particle densities for 17 ribbons grown from a single SiC-coated graphite die

one particle occurred after each 0.54 m of growth. For comparison, particle densities using graphite EFG dies are typically on the order of $5/cm^2$, and graphite CAST dies usually produce ribbons with less than 0.1 particles/cm^2. SiC particle densities of $< 3 \times 10^{-2}/cm^2$ are not unusual in the CAST growth configuration with graphite dies when a very high melt meniscus is maintained, and upon occasion a ribbon is grown with no particles. However, the large number of ribbons with zero or $\leq$2 particles seen with the SiC-coated die has not been encountered before with graphite dies. The difference in particle pick-up between graphite and SiC-coated graphite dies is not thought to be related to a difference in solubility. The dissolution[18] and solid contamination[7] characteristics of SiC and C in silicon have been shown to be similar.

In summary, of the materials studied, only SiC, graphite, vitreous carbon, boron carbide and CVD Si_3N_4 showed promise for capillary action silicon ribbon growth. Boron carbide exhibited excessive boron pick-up. CVD Si_3N_4 was not tested in an actual die configuration. Impervious SiC-coated dies gave the best results, although vitreous carbon and high density graphite are also very adequate for silicon ribbon growth.

III. Capillary Die Design

The height to which liquid will rise in a capillary is governed by Laplace's[16] equation which relates the pressure difference ΔP across a liquid surface to the two principal radii of curvature of the surface R_1 and R_2, and to the surface tension γ:

$$\Delta P = \gamma \left(\frac{1}{R_1} + \frac{1}{R_2}\right). \tag{1}$$

If liquid rises up a vertical capillary to a height h above the ambient liquid surface, then

$$\Delta P = \rho g h \tag{2}$$

where ρ is the liquid density (the density of the gas phase surrounding the liquid is assumed to be negligible) and g is the acceleration due to gravity. The capillary channel geometry and the wetting angle-θ of the liquid have an influence on R_1 and R_2. For a cylindrical capillary channel of radius r containing liquid which meets the wall at angle θ, it is typically assumed that the liquid surface is spherical so that $R_1 = R_2 = r/\cos\theta$. In this case the height of liquid rise is

$$h = \frac{2\gamma}{\rho g r} \cos\theta. \tag{3}$$

For a capillary formed between two vertical infinite plates separated by a distance s, it is usually assumed that the liquid surface is a segment of a cylinder. If the liquid meets the plates at angle θ, then $R_1 = s/2\cos\theta$ and $R_2 \rightarrow \infty$ so that

$$h = \frac{2\gamma}{\rho g s} \cos\theta. \tag{4}$$

This is also a good approximation for the rise height in the annular space between two concentric tubes, a configuration which may be used for the growth of tubular crystals. Fig. 11 shows h as a function of s for liquid silicon between plates having a wetting angle of 60 °. Plate separations of 0.03–0.05 cm are typical for silicon ribbon growth from graphite dies.

The die configuration of course must be such that the distance from the crucible melt level to the die-top, h_d, is less than the value of h given by Eq. (3) or (4). The advantage of making h_d large is that it is easier to separate thermal control at the die-top from thermal control in the crucible; the advantage to making the distance small is that a higher meniscus h_m between the die-top and the freezing interface is attainable. This decoupling between the die-top and the solid liquid interface is an important feature of the capillary action shaping technique. The reason that smaller values of h_d allow larger values of h_m is that the radius of curvature R_m of the meniscus surface in a vertical plane perpendicular to the crystal surface is, to a good approximation, given by

$$R_m = \gamma/\rho g h_d . \qquad (5)$$

Thus, R_m increases as h_d decreases. Since the meniscus base is pinned at the die edge (which has finite curvature on a microscopic scale), and the top of the meniscus must meet the crystal surface at a characteristic constant angle (11 ° for [111] silicon according to Surek[23]), a larger value of R_m allows a larger value of h_m. Of course, this arguement also implies that h_m increases as the die-top thickness increases. The maximum meniscus height occurs for infinetely thick dies not elevated above the melt surface i. e., for no die at all. This condition is approximated in the "web" region of the dendritic web and ESP growth processes mentioned previously.

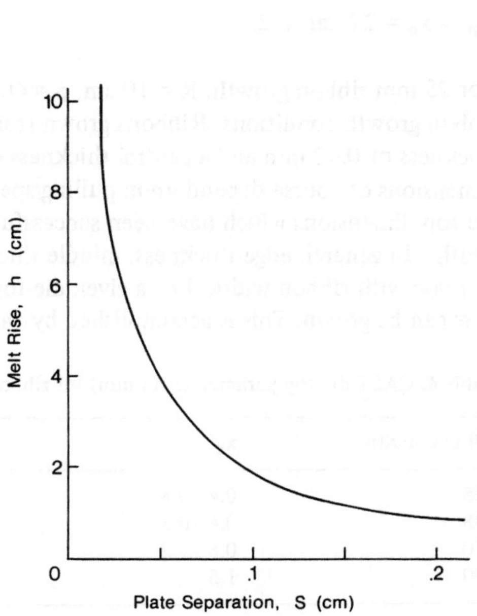

Fig. 11. Silicon melt rise between graphite capillary plates

A reasonable compromise between the extremes of a short meniscus remote from the crucible melt surface and a high meniscus with small h_d is provided by the die design of Fig. 1 with h_d in the range 2–3 cm. In this one-piece design, the melt raises up side channels and then flows across the top channel to feed the entire die-top. The die top surface is curved so that it is higher at the edges than in the middle. In this way, if the ribbon's solid-liquid interface is maintained approximately planar, then the interface is further from the die, at least in the central region. The central region is most critical for generation of defects in the ribbon. However, if the thickness of the die-top is kept constant while the die-top surface is curved, then a higher meniscus in the central region necessarily implies that the top of the meniscus is thinner there. This would cause the ribbon to be very non-uniform in thickness from one edge to the other (i. e., much thinner in the middle than at the ends). Thus, not only should the die top curve downward from the ends toward the central region, but it must also become thicker in the central region than at the ends, as in Fig. 1. The meniscus, then, has a wider base in the central region. The wider base, combined with the greater meniscus height in the central region, results in a more uniform ribbon thickness at the solid-liquid interface.

In summary, there are two things that are important to the die design: one is the curvature of the top surface, and the other is the thickening of the die-top in the central region. By proper choice of the curvature of the die top and the taper angle of the sides of the die, an optimum value for this variation of the thickness of the die top with position along the die can be obtained. R and ϕ (see Fig. 1) are chosen to optimize the values of $x_m - x_e$ and δ for successful ribbon growth. These parameters are given by:

$$\delta = R - \frac{W}{2 \tan\left(\sin^{-1} \frac{W}{2R}\right)} \qquad (6)$$

$$x_m - x_e = 2\,\delta \tan \phi/2 \qquad (7)$$

For 25 mm ribbon growth, R = 10 cm, $\phi = 40°$, and x_e = 0.40–0.80 mm give good ribbon growth conditions. Ribbons grown from such a die typically have an edge thickness of 0.42 mm and a central thickness of 0.21 mm for x_e = 0.55 mm. Exact dimensions of course depend upon pulling speed and other factors. Table 4 lists CAST die top dimensions which have been successfully used for various silicon ribbon widths. In general, edge thickness, middle thickness and deviation from flatness all increase with ribbon width. For a given die-top geometry, ribbons of different thickness can be grown. This is accomplished by varying the pulling speed and the thermal

Table 4. CAST die-top parameters (in mm) for ribbon widths of 25–100 mm

Ribbon width	x_e	x_m	σ
25	0.4–0.8	1.0–1.3	0.6–0.7
38	0.4–0.8	1.1–1.5	0.9–1.1
50	0.8–1.2	1.8–2.1	1.6–2.0
100	1.5	2.5	1.8

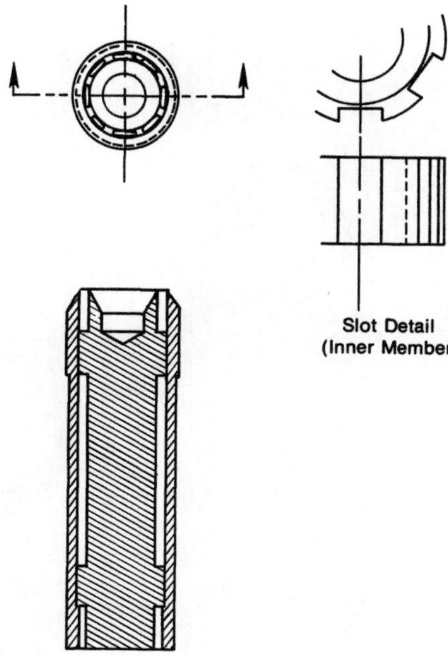

Fig. 12. A two-piece die for tube growth by the capillary action shaping technique

profile over the width of the ribbon growth front. Faster pull rates and hotter middle versus edge temperatures tend to decrease the thickness and vice versa.

A two-piece die with the same top configuration can also be used (Fig. 9). This design allows a more open channel which can be important for uniform dopant distribution in the ribbon. It also useful when coatings are to be applied to the entire die surface, such as CVD SiC coating of graphite dies.

The thin notches near the edges of the dies shown in Figs. 1 and 9 act as geometric limiters for the ribbon width. They avoid an abrupt thermal change near the ribbon edges such as would occur if the ribbon grew to the extreme edges of the die.

A two-piece die is particularly useful for tube growth, and a design example is shown in Fig. 12. The die consists of an outer tube surrounding an inner cylinder. Ribbed spacers on the inner cylinder maintain even spacing for the annular capillary channel.

Tube dies and one- and two-piece ribbon dies are all held in a one-piece graphite die holder machined to accept the die and hold it securely as shown in Figs. 4 and 13. The die-holder rests on a step in the cylindrical graphite susceptor. Side cut-outs in the die holder can be used for melt replenishment.

D. Silicon Ribbon Growth

I. Ribbon Growth Rates

An analysis of the heat balance at the growing interface for nondendritic crystal growth from the melt has been made by Billig[3]. His calculation indicated that the

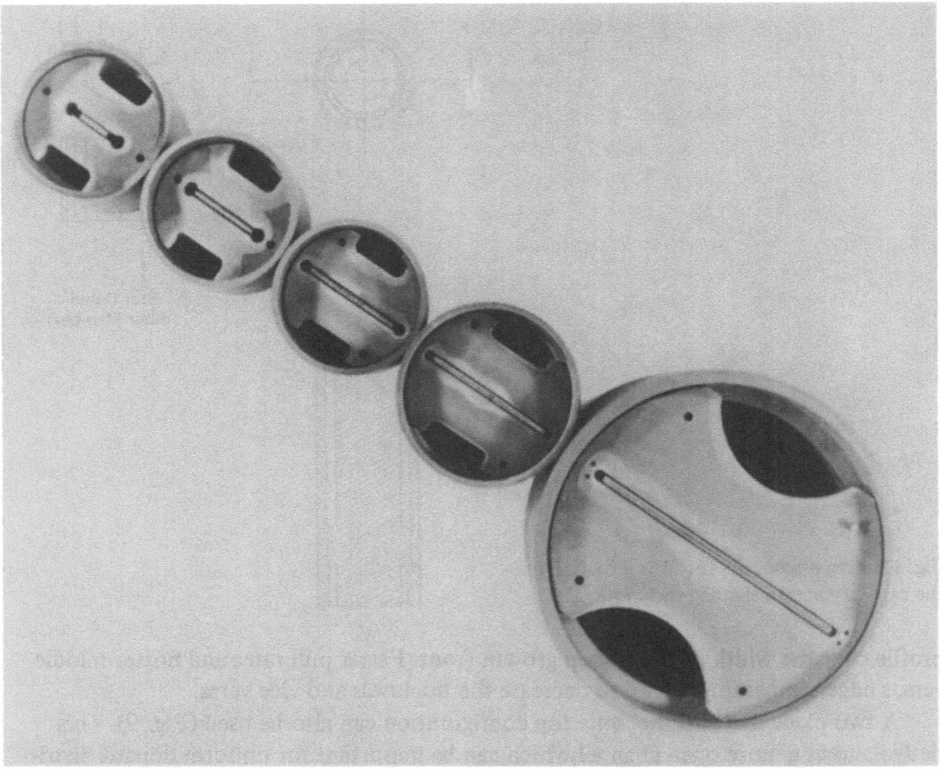

Fig. 13. Used dies and die holders for 12.5, 25, 38, 50, and 100-mm wide ribbon growth

maximum growth rate or pulling speed V_{max} would occur when the temperature gradient in the melt at the interface approached zero and would be given by

$$V_{max} = \frac{-K_m}{L\rho_m} \left(\frac{dT}{dx}\right)_m \tag{8}$$

where K_m is the thermal conductivity of the solid crystal at the melting temperature, L is the latent heat of fusion, ρ_m is the density of the crystal at the melting temperature, T is the crystal temperature, x is the distance along the crystal from the growing interface, and $(dT/dx)_m$ is the temperature gradient in the crystal at the growing interface.

A calculation of maximum ribbon growth rates has been carried out by Ciszek[9] under the following assumptions:

i) heat loss from the crystal is only by radiation to a cold environment,
ii) the solid liquid interface is planar,
iii) the thermal conductivity of the crystal varies inversely with temperature, and
iv) the surface emissivity ϵ of the crystal is temperature independent.

Heat from the solidification front is dissipated by conduction along the ribbon and radiation from the ribbon surface. Heat conduction along the ribbon is given by

$$Q = AK\frac{dT}{dx} = WtK\,\frac{dT}{dx}, \tag{9}$$

where A is the ribbon cross-sectional area, K is the thermal conductivity of the ribbon, W is the ribbon width, and t is the ribbon thickness. Heat loss from an element dS of the ribbon surface dx long via radiation to a cold environment is

$$dQ = \sigma\epsilon T^4\,ds = 2\,(W + t)\sigma\epsilon T^4\,dx. \tag{10}$$

For many materials of interest, particularly semiconductors such as silicon, the solid conductivity K can be approximately expressed as $K = K_m T_m/T$ [assumption (iii) above]. Making this substitution in Eq. (9), and combining Eqs. (9) and (10), we get the differential equation

$$\frac{d}{dx}\left(T^{-1}\,\frac{dT}{dx}\right) = \frac{2(W+t)\sigma\epsilon T^4}{WtK_m T_m} \tag{11}$$

The solution of Eq. (11) for boundary conditions

$$T = 0,\ \frac{dT}{dx} = 0 \quad \text{at } x = \infty$$

and

$$T = T_m \ \text{ at } x = 0$$

is

$$T = \left[2\left(\frac{(W+t)\sigma\epsilon}{WtK_m T_m}\right)^{1/2} x + T_m{}^{-2}\right]^{1/2} \tag{12}$$

The derivative of Eq. (12) evaluated at x = 0 is

$$\left(\frac{dT}{dx}\right)_{x=0} = \left(\frac{dT}{dx}\right)_m = -\left(\frac{\sigma\epsilon(W+t)T_m^5}{WtK_m}\right)^{1/2} \tag{13}$$

Combining Eqs. (8) and (13) results in the maximum growth rate under the stated assumptions:

$$V_{max} = \frac{1}{L\rho_m}\left(\frac{\sigma\epsilon\,(W+t)K_m T_m{}^5}{Wt}\right)^{1/2} \tag{14}$$

If the width-to-thickness ratio of the growing ribbon is large ($W/t >> 1$), Eq. (14) may be simplified to

$$V_{max} = (1/L\rho_m)\,(\sigma\epsilon K_m T_m{}^5/t)^{1/2}. \tag{15}$$

Even for $W/t = 10$, the maximum growth rate is within 5% of the rate for any greater W/t ratios. Thus, the maximum growth rate is essentially independent of width for wide ribbons and varies as $t^{-1/2}$.

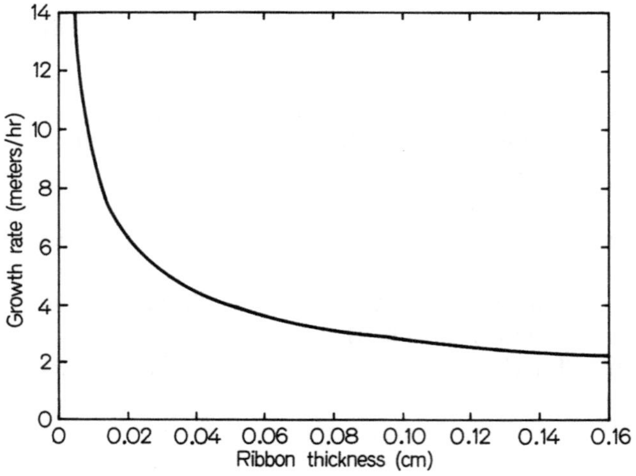

Fig. 14. Maximum growth rate vs thickness for silicon ribbon

The curve for maximum growth rate vs ribbon thickness from Eq. (15) is shown in Fig. 14 for the case of silicon.

Maximum Czochralski growth rates[9] are given by

$$V_{max} = (1/L\rho_m)(\sigma\epsilon K_m T_m^5/r)^{1/2}, \tag{16}$$

where r is the cylindrical crystal radius.

Although the maximum ribbon growth rates are considerably higher than maximum Czochralski growth rates, the amount of material produced per unit time is, of course, much smaller in the case of ribbons. One way of viewing this is to consider the rate of surface area produced during crystal growth, as is done in Fig. 15.

Actual growth rates for Czochralski crystals are about 0.2–0.3 V_{max} and for ribbons they are 0.3–0.4 V_{max}.

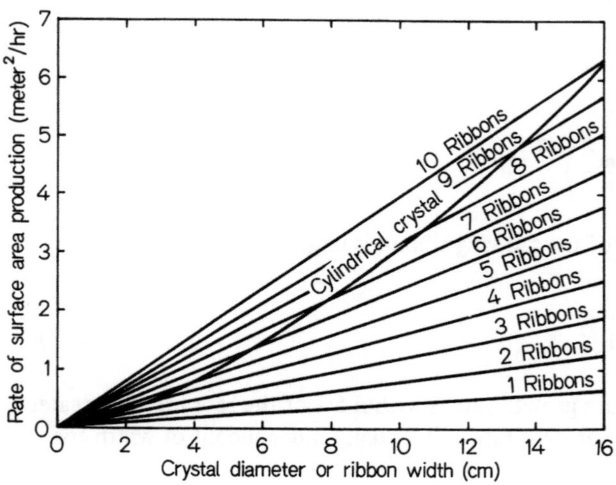

Fig. 15. Rate of surface-area production during growth vs Czochralski crystal dia. or ribbon width. A 0.05-cm ribbon thickness and postgrowth slicing of the Czochralski crystal into 0.05-mm thick wafers with 50% material loss is assumed. Maximum silicon growth rates are used. The effect of simultaneous multiple ribbon growth (up to 10 ribbons) is included

II. Thermal Geometry Control

Because of the relatively thin meniscus between the die-top and the freezing interface of the ribbon, thermal assymetries in the hot-zone can be disruptive to stable growth. This is particularly true as the ribbon width is increased. It is necessary to have some type of "on-line" thermal balancing and thermal profile control at the die-top. Some desireable features of such a thermal control system are that it be continuously variable, non-contaminating, and capable of heat extraction (cooling). This can be achieved with multiple inert-gas jets directed at various parts of the hot zone, each with an individually controlled flow adjustment. In Fig. 16 a system of four purge jets and flow-meters in a ribbon growth apparatus is shown. The two end jets impinge on the ribbon growth region at points 38 mm apart (corresponding to the edge positions of a 38-mm-wide ribbon). The two middle jets are directed so to be at either side of a ribbon being grown and pulled upward through the large central draw tube (which also serves as the exit tube for the inert gas-argon in this case). To use the system, more argon is directed to those parts of the hot-zone where the temperature is locally higher than desired. The cooling coefficient is about 6–7 °C/l/min at the ribbon edge. A change in argon flow directed at the edge will also have a small cooling effect at the middle of the ribbon.

Fig. 16. Inert-gas purge system for thermal geometry control; flow monitors are at upper left; purge tubes are at middle right

Figure 17 shows how assymetrical hot-zone profiles at the top of a 38 mm wide ribbon die can be flattened or symmetrically tailored by inert-gas thermal balancing. The system has provided the thermal control necessary for stable ribbon growth, and has allowed full-width seeding of the ribbons. The geometrical response of a growing, 38-mm-wide ribbon to step changes in argon flow directed at one of its edges has been checked. The growth conditions were first stabilized by adjusting the argon flows, as necessary, to attain full-width growth with no tendency for the ribbon to drift toward either edge of the capillary action shaping technique die. Then a step decrease in argon flow (directed at one edge of the ribbon) of 0.4 l/min was applied and caused the ribbon width to decrease (at the edge) at a rate of 8.7 mm/min per l/min, nearly linearly. A subsequent increase in flow by 0.4 l/min, back to the original level returned the ribbon to full width at a slower, non-linear average rate of 2.8 mm/min per l/min. The ribbon pulling speed was 14 mm/min. The geometry of the opposite ribbon edge was unaffected. Since the gas jets are directed at the growth region, it has been possible to use an open pulling port system and still retain clean ribbon surfaces of the type previously obtained only with a gas-tight system.

In addition to lateral thermal control at the growth front, it is also necessary to provide a vertical temperature distribution which allows rapid growth but does not induce stresses in the growing ribbon. From Eq. (8) it is evident that a steep temperature gradient is wanted near the interface for rapid growth. To achieve minimal stress levels, it is necessary and sufficient that ∇T be constant in the ribbon during cool down. A desireable approach is to maintain a high temperature gradient in the vertical direction near the die top ($\geqslant 50\ ^\circ C/mm$), changing to a low linear gradient ($\leqslant 5\ ^\circ C/mm$) in the ribbon region greater than about 1 cm above the interface. Ideally, gradients perpendicular to the pulling direction should be zero to satisfy ∇T = constant (in both magnitude and direction).

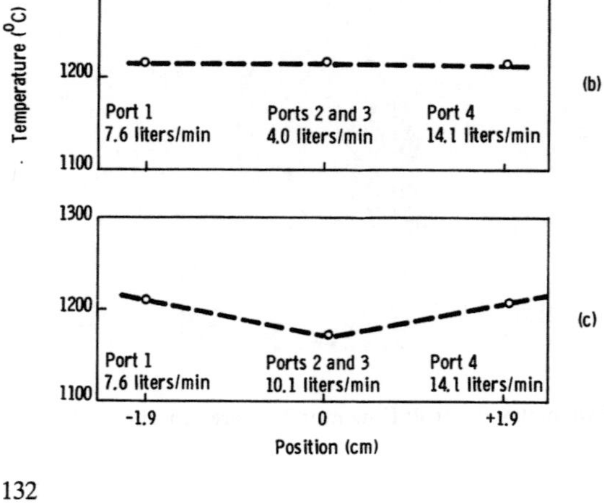

Fig. 17 a–c. An asymmetrical temperature profile (a) can be flattened (b) or symmetrically tailored (c) with the inert-gas system

III. Ribbon Surface Quality

Silicon carbide surface particles are the main detractors to ribbon surface quality, both by their physical presence on the surface and because of the longitudinal troughs or tracks they leave on the ribbon when they are present in the liquid meniscus at the die top. Another major contribution to roughness consists of sporadic, heavy horizontal striations which can arise when the solid-liquid interface is in very close proximity to the die top. As can be seen from Fig. 18, die design can greatly reduce the severity of these two sources of roughness. A close look at Fig. 18 (b) shows that roughness from grain boundaries and fine horizontal interface striations still remains with the improved die design.

The transverse interface striations are quite uniformly spaced and, under certain lighting conditions, can act as a diffraction grating causing the ribbon surface to appear multicolored. In fact, a spectrum can be projected onto a white surface by using a ribbon as a grating. The striation line density was measured for nine ribbons which were grown at rates from 15 to 31 mm/min (see Table 5). Line densities ranged from 620–970 lines/cm. The product of the line density and the growth rate yields the striation frequency, which is also given in Table 5. Line density and frequency do not change appreciably with distance along the ribbon, as can be seen from the data for ribbon 51012. Several explanations of these striations have been proffered, including

a) AC modulation of the rf generator output power,

b) room vibration due to heavy equipment motors (vacuum pumps, air conditioning motors, etc.) running at about 1725 RPM (28.8 cycles/s), and

c) oscillations associated with surface tension driven flow in the meniscus as proposed by Schwabe and Scharmann[19)]. No conclusive cause has been identified.

To determine whether or not the surface striations correlate with dopant distribution striations, a ribbon was beveled at a 3.5 ° angle (top, Fig. 19) so that spreading resistance measurements could be made at 2.5 μm intervals, on the lapped surface, in a direction perpendicular on the striation lines. The striation lines, in this case, were spaced at approximately 10 μm intervals and had a nearly sinusoidal peak-to-valley undulation with an amplitude of about 0.37 μm (as calculated from the bevel angle and the structure at the bevel/surface intersection in Fig. 19). Resistance fluctuations of up to about 12% are evident, but do not appear to correlate with the surface striations.

Another detractor from surface quality is the formation of surface films on the ribbons due to the presence of quartz and graphite in the growth furnace and their reaction with silicon. In the ribbon apparatus at growth temperature (1450 °C), the primary hot zone materials are C, Si (liq.) and SiO_2. The reactions

$$2Si + SiO_2 \longrightarrow 2SiO, \; C + SiO_2 \longrightarrow SiO + CO,$$

and

$$3C + SiO_2 \longrightarrow SiC + 2CO$$

are feasibly operative. Thus there are a number of means by which foreign materials may be transported via the vapor phase to the ribbon surface.

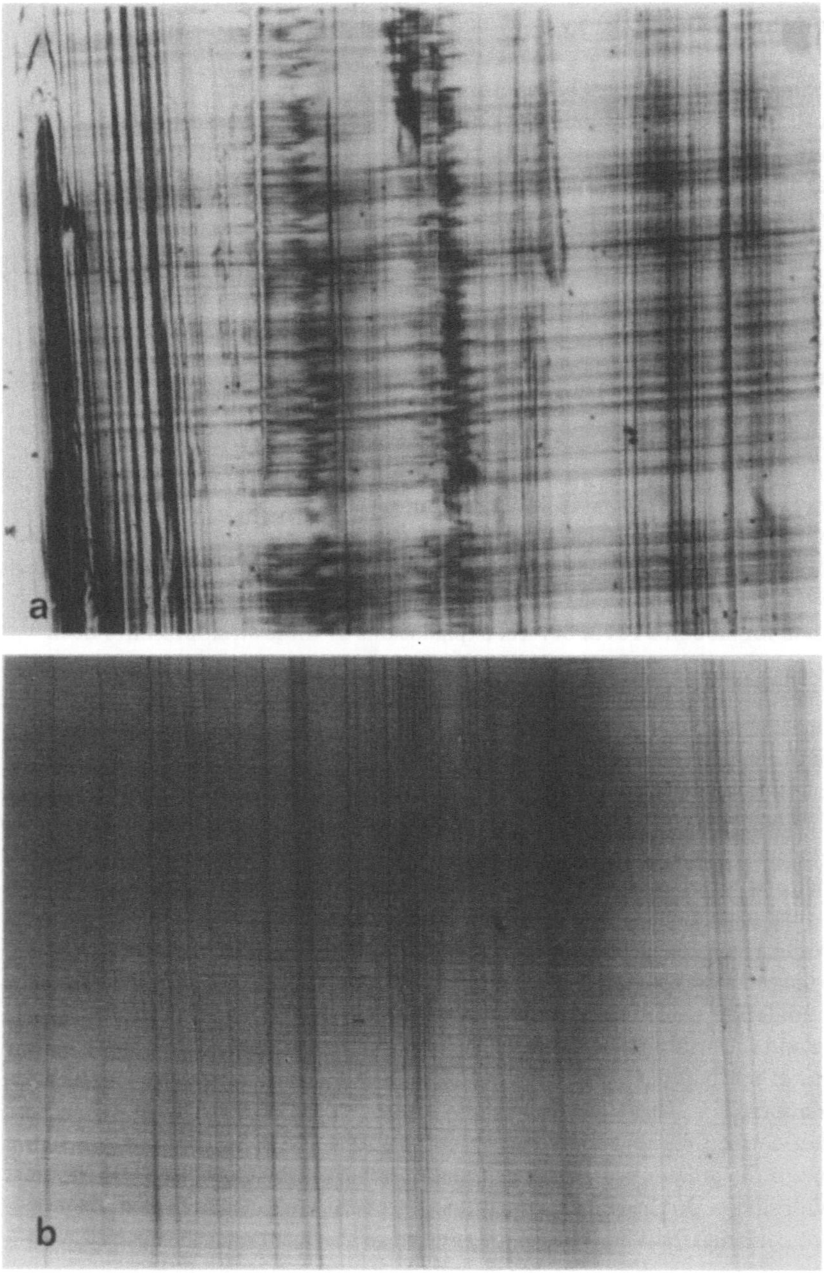

Fig. 18. 21X micrographs of as-grown ribbon surface; (a) EFG die design and (b) CAST die design

In the growth system of Fig. 4, the inert gas inlet is typically at the bottom of the furnace and the gas flows upward past the melt region, exiting through an oval

Table 5. Growth rates, solid liquid interface striation line densities and line frequencies for ten ribbons

Ribbon Number	Distance from Seed (cm)	Growth Rate (cm/min)	Line Density (lines/cm)	Frequency (lines/s)
51007	65	1.5	756	18.9
51012	12	1.9	960	30.4
51012	100	1.9	970	30.7
51014	65	1.7	956	27.1
51017	32	3.1	660	34.1
51019	64	2.9	636	30.7
51022	66	2.0	940	31.3
51023	69	2.2	800	29.3
51027	66	2.5	620	25.8
51107	57	1.5	780	19.5

tube at the furnace top. The ribbon is also pulled through this tube. In an attempt to eliminate ribbon surface film formation, a closed (gas-tight) system modification was made to the ribbon puller, and the inert argon purge system was altered to direct the purge gas at the ribbon/melt interface before exiting at the bottom of the furnace. This resulted in a cleaner appearance of the melt at the die top and also a cleaner looking ribbon surface. However, there is evidence from TEM examinations of SiC film formation near the seed interface. The film growth appears to be by chemical vapor deposition during the relatively long time (approximately five minutes) the seed remains stationary at the die top. A similar film is seen on the ribbon, if growth

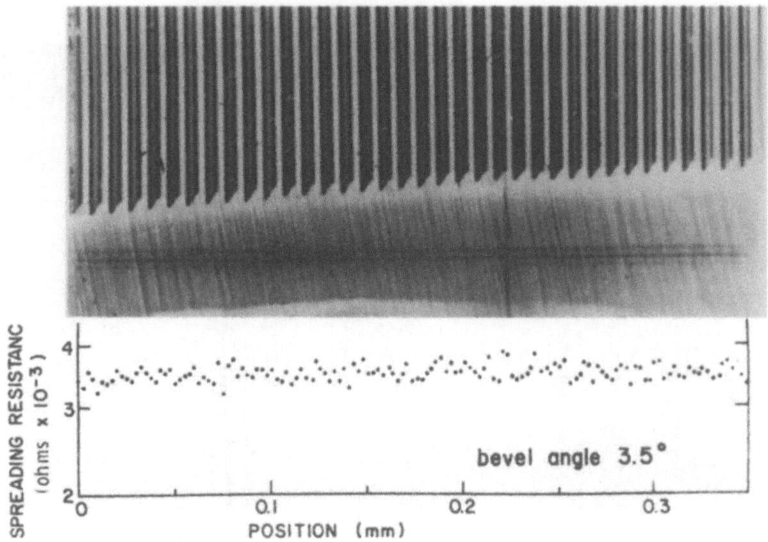

Fig. 19. Spreading-resistance measurements, on a 3.5 ° beveled surface, perpendicular to surface striations

is halted for several minutes with the ribbon remaining at the die top. During growth at typical speeds, the film is not noticeable, but is still diminutively present.

This type of film thus is dependent upon the residence time of the ribbon surface near the hot zone components. Upon microscopic examination, the film is seen to consist of dendritic structures about 1–10 μm in size. Their spacing increases, and their size decreases for shorter residence times. Electron diffraction patterns indicate that the film consists of randomly oriented β-SiC dendritic crystals, which occur either randomly over the ribbon surface or preferentially along grain boundaries. The surface dendrites have been found to influence generation lifetime of the silicon ribbons, and contribute to the occurrence of leaky MOS dots[20]. However, etching the surface before processing eliminates this problem. The most serious effect of the SiC growth is its interference with controlled seeding at the onset of ribbon growth.

Because of vapor phase silicon carbide formation during seeding, the original seed/crystal relation is rapidly lost and the final ribbon surface structure is mostly dominated by twin lines in the <112> direction. All inclined twins grow out relatively fast. If the pick-up of silicon carbide particles from the die top can be avoided, steady-state growth is dominated by surface orientations close to a <110> direction.

IV. Meniscus Effects

As was mentioned in the section on die design, the meniscus height is intimately linked to ribbon surface properties and structural perfection. It is helpful to have some quantitative means of determining the meniscus height, h_m. A photographic technique is described below.

The meniscus is photographed at a known arbitrary viewing angle, θ, relative to the horizontal plane. For example, the meniscus of Fig. 20 was photographed at an angle of 52–1/2 °. The distance from the seed to the solid-liquid interface at the

Fig. 20. Meniscus photograph of 50 mm wide ribbon during growth at 19 mm/min. The meniscus is the dark thin crescent. The oval openings in the heat shields near the interface are reflected in the shiny ribbon surface

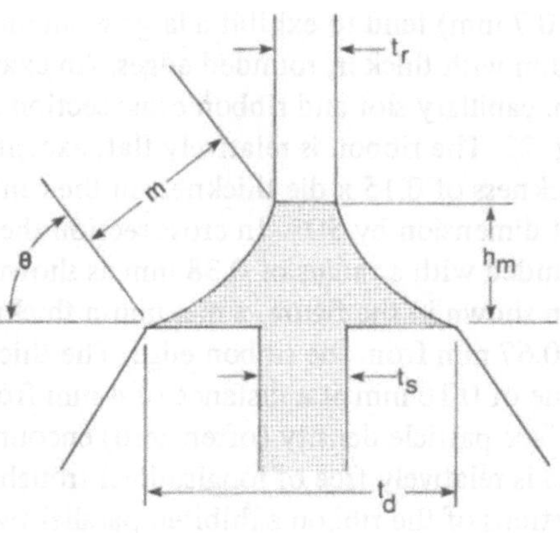

Fig. 21. Meniscus configuration in CAST ribbons

time of the photograph is noted. Then, after growth, the ribbon thickness, t_r, at the location photographed is measured. The die-top thickness, t_d, is measured prior to the growth run. The apparent meniscus height, m, is obtained from a densitometer scan of the photographic negative using a double-beam recording microdensitometer.

Once m, θ, t_r, and t_d are known, the meniscus height, h_m, can be obtained from

$$h_m = \frac{m}{\cos\theta} - \frac{t_d - t_r}{2} \tan\theta. \quad (17)$$

The geometrical relationship of these quantities is shown in Fig. 21. The above analysis assumes that the ribbon grows symmetrically centered at the die top as shown in Fig. 21. If this is not the case, the meniscus should be photographed from both sides. Two values of the apparent meniscus height m_1 and m_2 are thus obtained. It can be mathematically shown that the correct meniscus heigth is then given by

$$h_m = \frac{(m_1 + m_2)}{2\cos\theta} - \frac{t_d - t_r}{2} \tan\theta. \quad (18)$$

Meniscus heights in the capillary action shaping technique typically range from 0.5 to 0.8 mm for 50 mm wide silicon ribbons. Those grown with a high melt meniscus

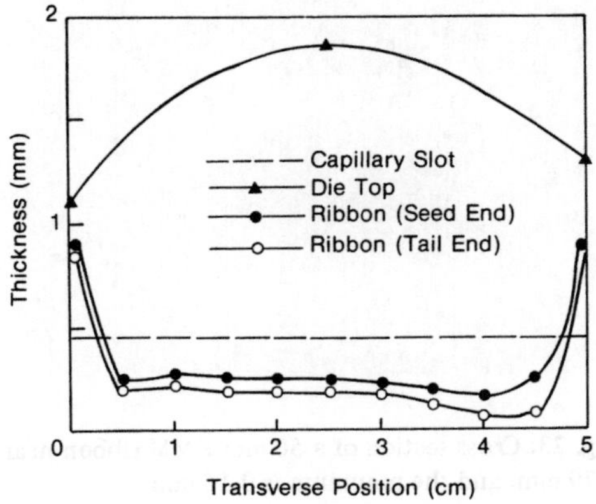

Fig. 22. Thickness of die top, capillary slot, and ribbon as a function of transverse position for a 50 mm wide ribbon grown in the high melt meniscus (HMM) mode

(> 0.7 mm) tend to exhibit a large grain morphology and to be thin in the central region with thicker, rounded edges. An example of the relative dimensions of the die top, capillary slot and ribbon cross-section for high melt meniscus growth is shown in Fig. 22. The ribbon is relatively flat, except for several mm near each edge, and has a thickness of 0.15 x die thickness in the central region. It is thinner than the capillary slot dimension by 50%. In cross section the extreme edges of the ribbon appear rounded with a radius of 0.38 mm as shown in Fig. 23. At the cross-sectional position shown in the figure, a maximum thickness of 0.79 mm is reached at a distance of 0.67 mm from the ribbon edge. The thickness gradually decreases to a uniform value of 0.16 mm at a distance of 4 mm from the edge. Because of the very low SiC surface particle density (often zero) encountered with HMM growth, the ribbon surface is relatively free of longitudinal troughs often associated with such particles. Portions of the ribbon exhibited parallel twinning as shown in the central ribbon region of Fig. 24. However, much of the surface exhibited large single grains of the

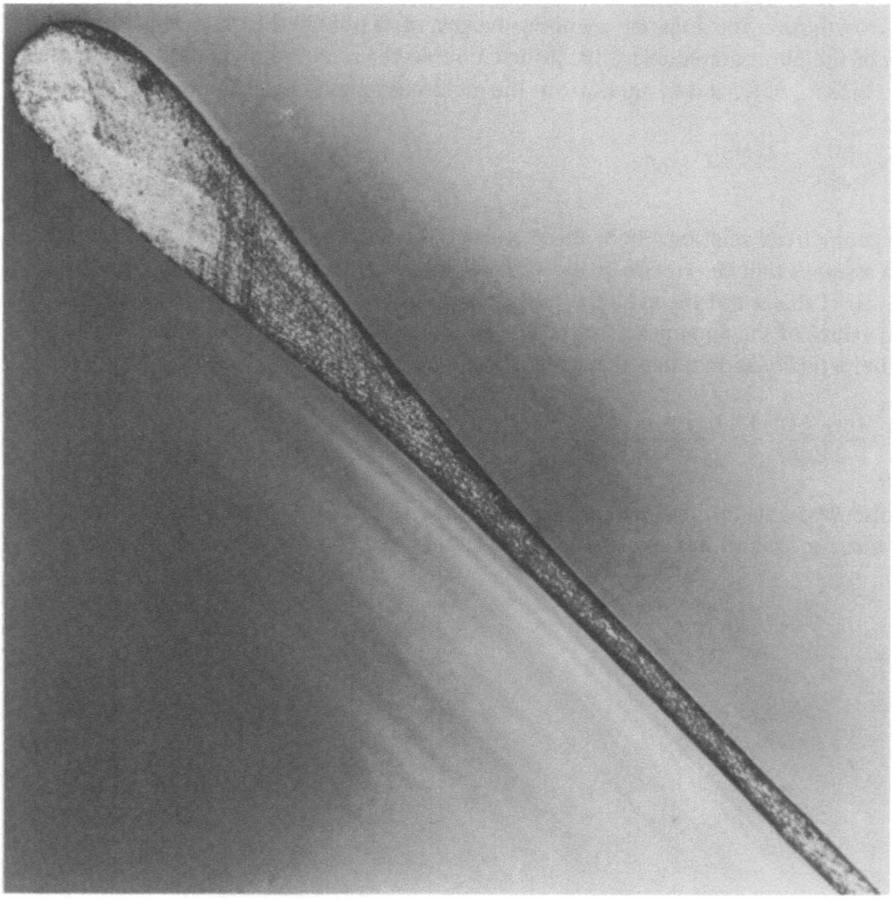

Fig. 23. Cross section of a 50 mm HMM ribbon near an edge. The maximum thickness is 0.79 mm, and the minimum is 0.15 mm

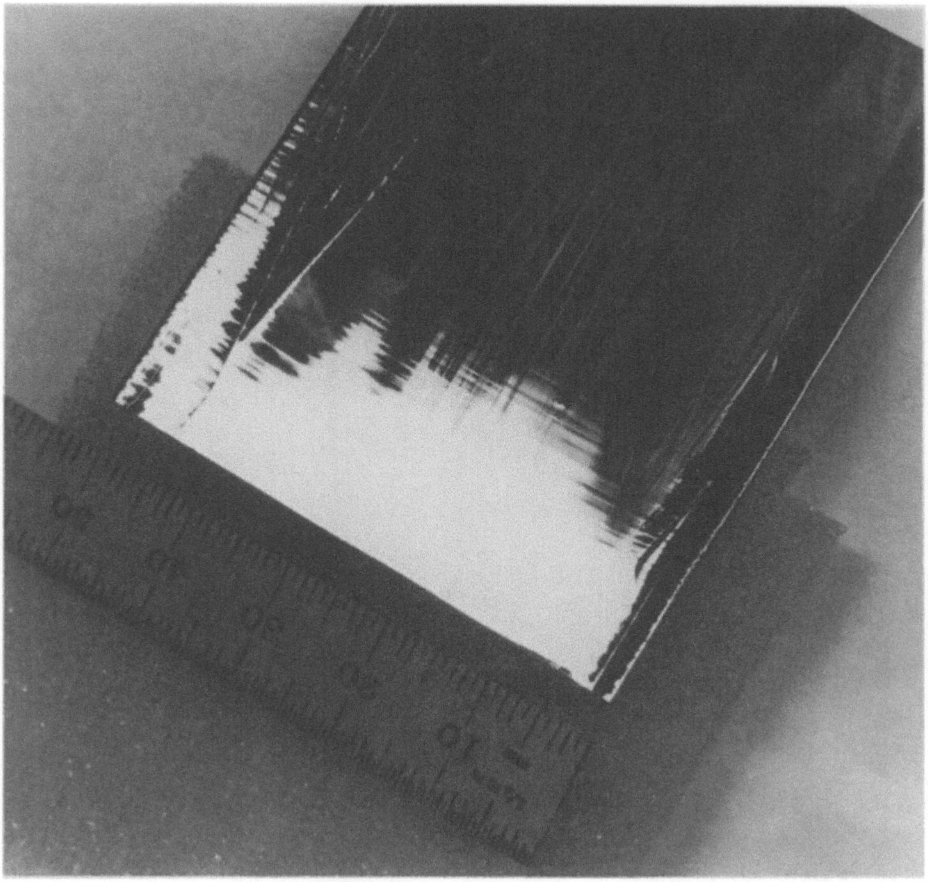

Fig. 24. Surface of a 50 mm wide HMM ribbon

order of 8 mm in width. Frequently these grains were separated by twins with a raised surface morphology at the twin-ribbon surface intersection. Examples can be seen near the edge of the ribbon in Fig. 24. A 1.2 meter long by 50 mm wide silicon ribbon grown by the HMM capillary action shaping technique is shown in Fig. 25.

E. Silicon Tube Growth

Silicon tubes can be grown by capillary action shaping (Ciszek[8]) using the same apparatus employed for ribbon growth (except of course for the die and heat shields). The die has been described in Fig. 12. The growth technique is shown schematically in Fig. 26.

A 5 cm diameter quartz crucible resting in a conductive rf heated susceptor held a 50 g silicon charge. The high density graphite capillary die was immersed in the

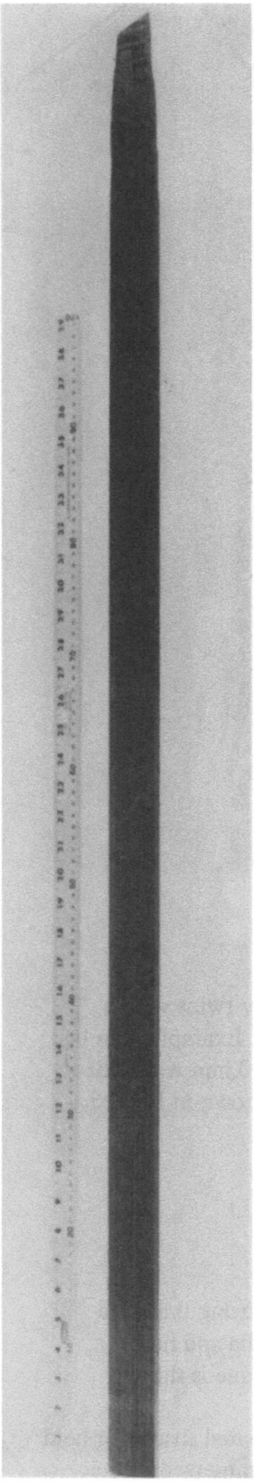

Fig. 25. A 50 mm wide by 1.2 m long silicon ribbon grown by the Capillary Action Shaping Technique

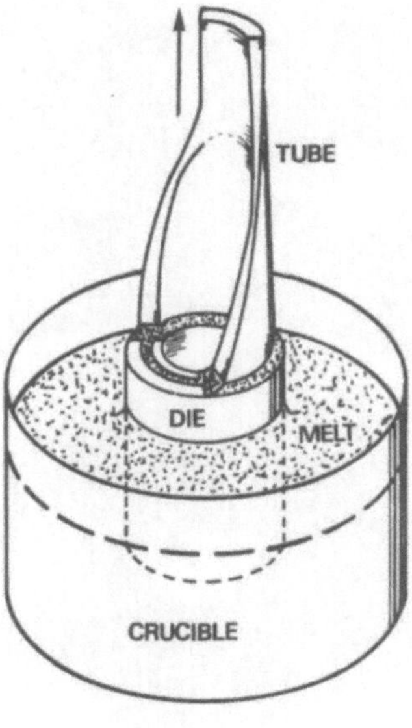

Fig. 26. A schematic diagram of capillary action tube growth

liquid silicon, and supported by a holding plate (not shown in Fig. 26) slightly below its top surface. The die top was 7 mm outer diameter by 4.4 mm inner diameter with a 0.64 mm thick capillary slot. Two horizontal heat shields were concentrically placed above the die and holder; the upper shield is visible in Fig. 27.

Thin <111> axis seeds were used to initiate growth at some point of the capillary slot (Fig. 27a). The system temperature (as monitored by an optical radiation pyrometer with 0 to 10 mV voltage output) was then lowered while simultaneously increasing the upward pulling speed of the seed lifting mechanism. The growth was thus caused to spread circumferentially around the die top (Fig. 27b). Further reduction of the system temperature allowed increased spreading of the growth (Fig. 27c) until the tube walls closed. Equilibrium tube growth is pictured in Fig. 27d. Tubes were normally closed in a distance of 5 to 15 cm beyond the seed, and were grown in closed lengths of over 1 m. Growth rates were varied from 11 to 21 mm/min and had an influence on tube diameter and wall thickness, as did the system temperature.

Both growth rate and system temperature were investigated as parameters for tube diameter and wall thickness control. The temperature dependence was determined by growing a tube and gradually increasing the temperature as growth progressed beyond the point of closure. Over a growth distance of 1.03 m, the temperature was increased 6.7 °C while keeping growth speed constant at 11 ± 1 mm/min. The tube was then sectioned at six positions corresponding to six different values of the system temperature.

Analysis of the profiles led to the temperature dependence of average outside diameter, inside diameter, and wall thickness depicted in Fig. 28. The dependence is approximately linear in each case, with the respective magnitudes of −0.13 mm/°C, 0.08 mm/°C, and −0.105 mm/°C. Increasing the speed of growth while maintaining a constant system temperature also causes a monotonic decrease in tube wall thickness.

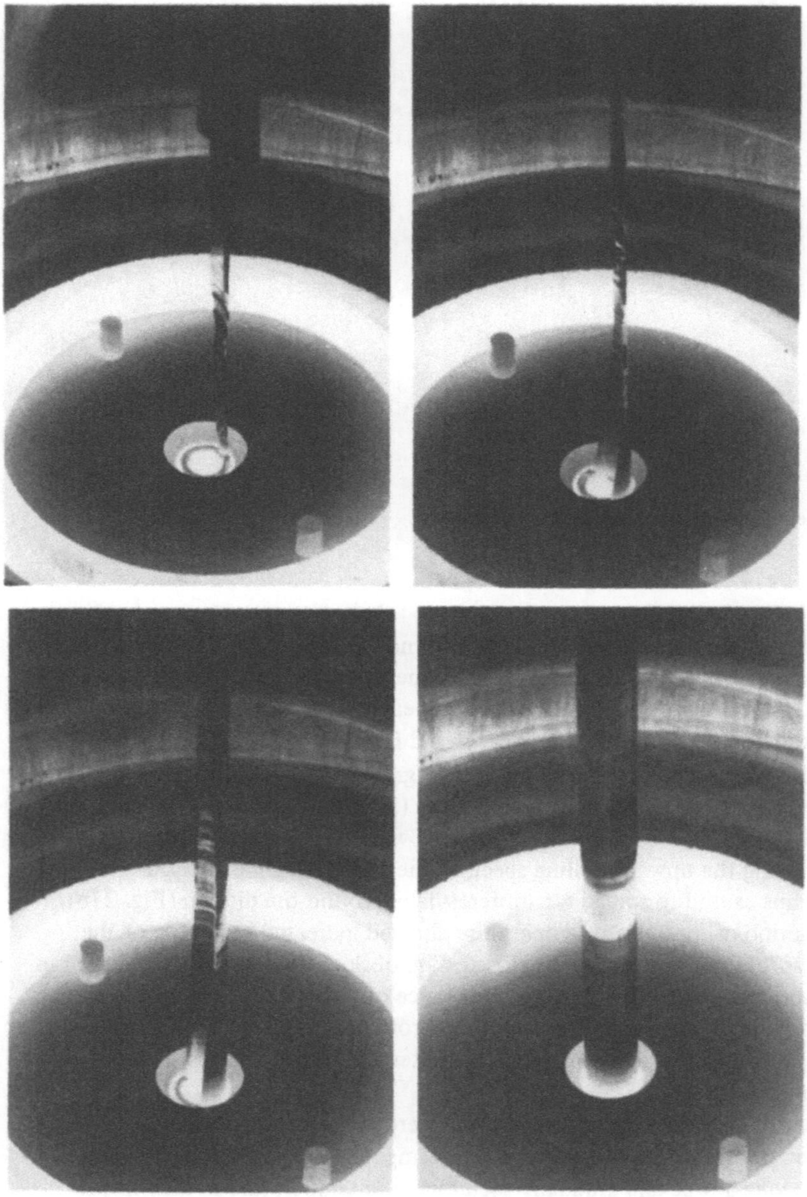

Fig. 27 a–d. Stages in the growth of a silicon tube; (a) Seeding, (b) Circumferential spreading by temperature reduction and pulling rate increase, (c) Further spreading by temperature reduction, (d) Closed, equilibrium growth

The decrease in tube wall thickness is of course associated with a higher melt meniscus. As in the case of ribbon growth, the tube crystallographic properties are improved with higher melt meniscus (HMM) growth. The etched tube cross-sections of Fig. 29 compare the grain structure in two tubes where the ratio of tube-wall to die-

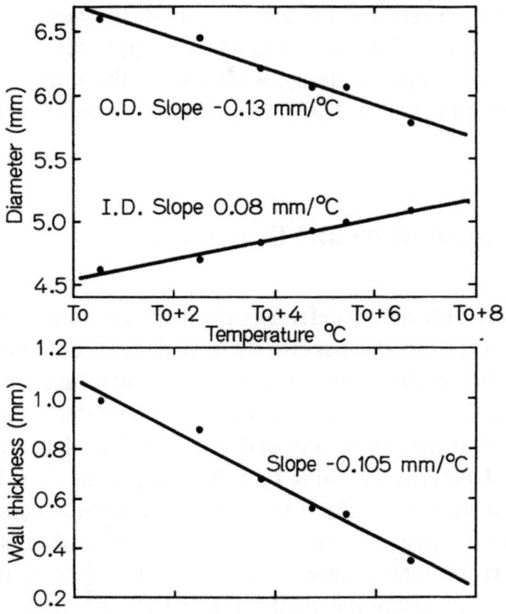

Fig. 28. Dependence of (a) average outside diameter and inside diameter and (b) wall thickness upon relative system temperature

wall thicknesses is 0.77 and 0.43 respectively. In the latter one, a single grain extending over about 1/5 of the tube-wall can be seen. The defect structure in this tube is dominated by linear boundaries, whereas the thicker tube (lower meniscus height) contains a number of irregular boundaries in addition to the planar ones.

Tubes appear to be somewhat easier to grow than ribbons, in that growth is stable over a wider range of system temperatures and growth rates. However, changes

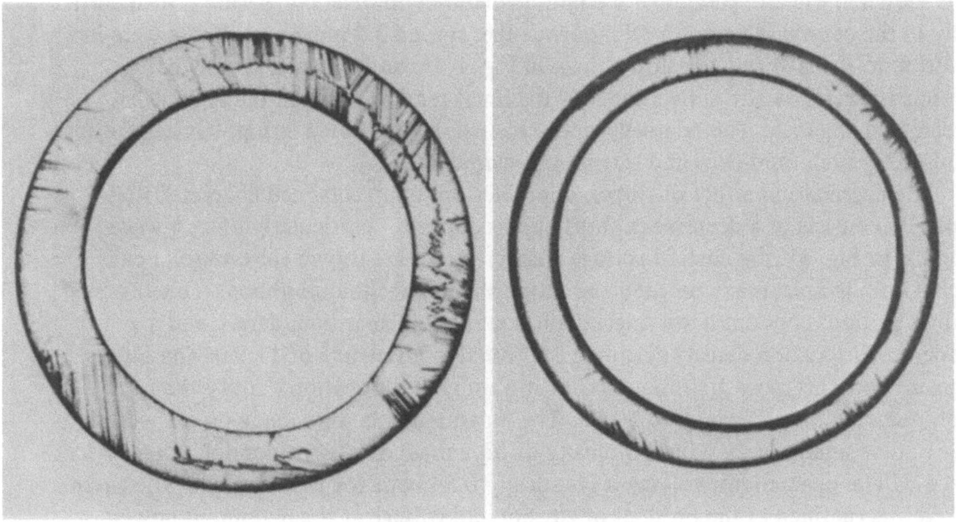

Fig. 29. Etched cross section of tubes. Ratio of tube-wall to die-top thickness is 0.77 (*left*, low meniscus) and 0.43 (*right*, high meniscus)

in growth rate and temperature affect the tube dimensions and, in fact, can be used to obtain various dimensions from a given die.

While one die design was used for the work reported here, it should be possible to make tubes of smaller and larger sizes by suitable die design changes.

F. Applications and Discussion

Electrically active defects in capillary action grown ribbons have limited their usefulness for most devices currently made with dislocation-free silicon. However, photovoltaic devices made on these ribbons can function with reasonably good efficiencies. Still, the cell efficiency is related to the defect types and densities in the material. Some of the electrically active defects have been mentioned. It is possible to classify the defects as was done by Schwuttke, et al.[20] using MOS-Ct measurements. Strongly active defects include silicon carbide dendrites and particulate inclusions, grain boundaries, degenerated twins, dislocation bands, and dislocation densities above $10^6/cm^2$. Some moderately active defects are multiple, overlapping stacking faults, dislocation densities above $10^4/cm^2$, and non-coherent twins. Relatively inactive defects are those involving coherent (first order) twins, stacking faults, and dislocation densities below $10^4/cm^2$.

Impurity effects have not been discussed in detail, but it must be remembered that impurity interactions with crystallographic defects also have an influence on electrical transport properties. Spreading resistance measurements show that planar defect boundaries usually do not perturb the electrical conductivity, while some irregular grain boundaries are associated with abrupt changes in the concentration of doping impurities. Capillary channel design also plays a role in impurity distribution, as is shown in Fig. 30. Here, five 0.8 mm diameter cylindrical holes were drilled vertically in the central region of a 38-mm-wide die, spaced 3.2 mm apart. These were in addition to the side and top slots shown in Fig. 1. It can be seen that resistivity minima (net electrically active impurity maxima) tend to occur in the regions between feed channels. The relatively open-channel design shown in Fig. 9 is helpful for minimizing such fluid-flow and stagnation-related effects.

The degradation effect of ribbon imperfections upon solar cell characteristics typically consists of a decrease in short-circuit current -- particularly at long wavelengths. In Fig. 31, the current-voltage characteristics are shown for two solar cells, each 4 cm^2 in area, made on capillary action shaping technique ribbons. One of the ribbon sections contains a low-angle grain boundary, planar boundaries, and a moderate dislocation density, yielding a conversion efficiency of 11.9%. The other segment contains more grain boundaries and a higher dislocation density than the first, reducing the efficiency to 10.8%. The measurements were made on a Spectrolab X25L solar simulator, at a light intensity of 83.2 mw/cm^2, corresponding roughly to AM 1.5. The open circuit voltage is identically 0.54 volts for the two cells. The curve fill factors (defined as the product of current and voltage at the maximum power point divided by the product of short-circuit current and open-circuit voltage) are also the same for both cells at 0.77. Thus, the primary characteristic difference be-

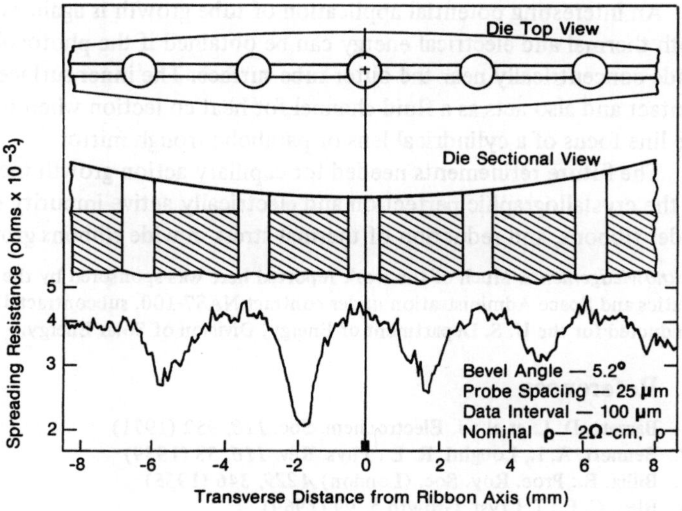

Fig. 30. Central portion of a CAST die, showing effect of capillary channels upon dopant distribution

tween cells is the short-circuit current density which is 23.6 ma/cm^2 for the 11.9% cell and 21.5 ma/cm^2 for the 10.8% efficient cell. The primary loss in short-circuit current occurs at long wavelengths (greater that 0.6 μm). The peak response is about 0.53 ma/mw near 0.8 μm for good CAST cells, dropping off more than 50% for wavelengths shorter than 0.48 micrometers or longer than 1.04 μm. For worst-case CAST cells, the peak response is 0.37 ma/mw near 0.69 μm, dropping off more than 50% for wavelengths less than 0.44 μm or greater than 0.92 μm.

The primary reasons for pursuing this type of ribbon as a solar cell material are elimination of sawing and associated material loss, high linear growth rates, and decoupling of thermal control at the die-top from that at the crucible. Unfortunately, throughput is low in spite of the fast pulling speeds, as was shown in Fig. 15. Another negative feature of this process is the stringent thermal control requirement near the solid-liquid interface. Hence, it is not clear at this stage whether there are net advantages of capillary action growth over other growth techniques.

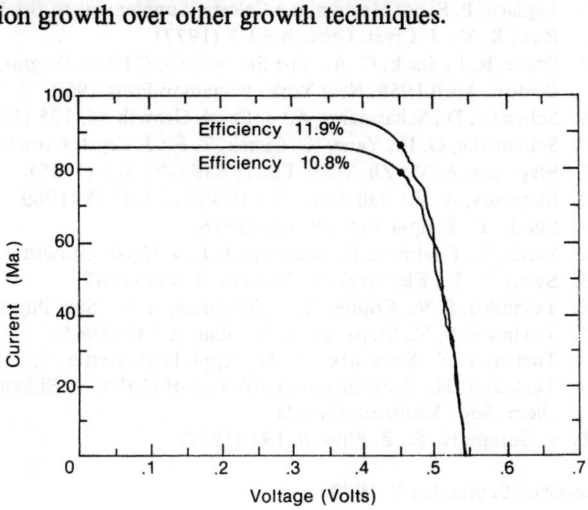

Fig. 31. Current/voltage characteristics for two solar cells made on 4 cm^2 silicon ribbons having different defect densities

An interesting potential application of tube growth is again as a solar collector. Both thermal and electrical energy can be obtained if the photovoltaic junction is made concentrically near the outer tube surface. The inner surface receives the back contact and also acts as a fluid channel for heat collection when the tube is placed at the line focus of a cylindrical lens or parabolic trough mirror.

The future refinements needed for capillary action growth include improvement of the crystallographic perfection and electrically active impurity control, growth of wider ribbons, and reduction of thermal stress in wide ribbons grown at high speeds.

Acknowledgements. Much of the work reported here was sponsored by the U. S. National Aeronautics and Space Administration under contract NAS7-100, subcontract JPL 954144, and was conducted for the U. S. Department of Energy, Division of Solar Energy.

G. References

1. Barrett, D. L. et al.: J. Electrochem. Soc. *118,* 952 (1971)
2. Bennett, A. I., Longini, R. L.: Phys. Rev. *116,* 53 (1959)
3. Billig, E.: Proc. Roy. Soc. (London) *A229,* 346 (1955)
4. Bleil, C. E.: J. Cryst. Growth *5,* 99 (1969)
5. Boatman, J., Goundry, P.: Electrochem. Technol. *5,* 98 (1967)
6. Brissot, J. J., Raynaud, H.: Electrochem. Technol. *1,* 304 (1963)
7. Ciszek, T. F.: Mater. Res. Bull. *7,* 731 (1972)
8. Ciszek, T. F.: Phys. Stat. Sol. (a) *32,* 521 (1975)
9. Ciszek, T. F.: J. Appl. Phys. *47,* 440 (1976)

9a. Ciszek T. F., Hurd, J. L.: "Melt Growth of Silicon Sheets by Edge-Supported Pulling". Proceedings of the Symposia on Electronic and Optical Properties of Polycrystalline or Impure Semiconductors and Novel Silicon Growth Methods. Ed. by K. V. Ravi and B. O'Mara. St. Louis, MO; 11–16 May, 1980. The Electrochemical Soc., Pennington, NJ; Proceedings Volume 80–5; 213–222

10. Ciszek, T. F., Schwuttke, G. H.: Phys. Stat. Sol. (a) *27,* 231 (1975)

10a. Ciszek, T. F. et al.: Proc. 15th IEEE Photovoltaic Specialists Conf. (1981)

11. Ciszek, T. F., Schwuttke, G. H., Yang, K. H.: J. Cryst. Growth *46,* 527 (1979)
12. Dermatis, S. N., Faust, J. W., Jr.: IEEE Trans. Commun. and Electronics *82,* 94 (1963)
13. Gurtler, R. W. et al.: J. Electronic Materials 7, 441 (1978)
14. Koyanagi, T.: Proc. 12th IEEE Photovoltaic Specialists Conf., 627 (1976)
15. LaBelle, H. E., Jr.: Mater. Res. Bull. *6,* 581 (1971)
16. Laplace, P. S. de: Mechanique Celeste, Supplement to Book 10 (1806)
17. Ravi, K. V.: J. Cryst. Growth *39,* 1 (1977)
18. Scace, R. I., Slack, G. A.: The Si-C and Ge-C Phase Diagrams, Proc. Conf., Silicon Carbide, Boston, April 1959, New York: Pergamon Press 1960
19. Schwabe, D., Scharmann, A.: J. Cryst. Growth *46,* 125 (1979)
20. Schwuttke, G. H., Yang, K., Ciszek, T. F.: J. Cryst. Growth *43,* 329 (1978)
21. Stepanov, A. V.: Zh. Tekh. Fiz. (USSR) *29,* 381 (1959)
22. Stepanov, A. V.: Bull Acad. Sci. (USSR) *33,* 1775 (1969)
23. Surek, T.: Scripta Met. *10,* 425 (1976)
24. Surek, T., Chalmers, B., Mlavsky, A. I.: J. Cryst. Growth *42,* 453, (1977)
25. Swets, D. E.: Electrochem. Technol. *5,* 385 (1967)
26. Tsivinskii, S. V., Koptev, Y. I., Stepanov, A. V.: Sov. Physics-Solid State *8,* 449 (1966)
27. Tsivinskii, S. V., Stepanov, A. V.: ibid. *7,* 148 (1965)
28. Tucker, T. N., Schwuttke, G. H.: Appl. Phys. Letters *9,* 219 (1966)
29. Tucker, T. N., Distribution Coefficient of Carbon in Silicon, paper Oct. 1968 Meet. Electrochem. Soc., Montreal, Canada
30. v. Gompertz, E.: Z. Phys. *8,* 194 (1922)

Received September 7, 1979

Crystal Growth of Silicon Ribbons
For Terrestrial Solar Cells by the EFG Method

Fritz V. Wald

Mobil Tyco Solar Energy Corporation, 16 Hickory Drive, Waltham, Massachusetts 02154, USA

Crystals: Growth, Properties and Applications 5
© Springer-Verlag Berlin Heidelberg 1981

1 Introduction

The growth of shaped silicon crystals has been attractive for a number of years and several methods to obtain various shapes, in particular flat ribbons, have been developed. Recently, however, the problem has been much more sharply focused due to the desire for the preparation of very low cost solar cells for terrestrial power generation, and a number of articles in the present volume address this problem.

It is easy to recognize intuitively in this context the fact that any method used for such a purpose would probably have to have capabilities such as high volume output in ribbon form, i.e., high growth rates, and ease in growing wide ribbon in a continuous mode with little or no operator attention. Also, simple calculations[1] can readily show that one crystal growth operator must be capable of supervising the growth of a large number of ribbons, or the total ribbon cost easily becomes dominated by the cost of operating labor.

At the time when experimentation on the growth of silicon ribbon by EFG (Edge Defined Film Fed Growth) started[2–4], we could already look back at significant operating experience in the growth of sapphire by this method[5–7] which by then was used for the commercial production of various sapphire shapes. From these results, it was but a small step to conclude that the growth of silicon ribbons by this technique could lead to significant cost reductions if they could be grown under conditions similar to sapphire, and if under these conditions their semiconductor properties would not be compromised to such an extent that reasonable device efficiencies could no longer be realized.

This paper then is not a description of the historical evolution of the method or an analysis in detail of the principles on which it is based, because these have been amply covered in the literature. Rather, it is intended as a general review of its current status which proceeds from an abbreviated discussion of the fundamental principles directly to questions of speeds, geometrical tolerances and other practical growth variables, which can affect ribbon output in production situations. Machine designs will also be covered. They are believed to contain all the concepts that make it possible to fabricate ribbons in volume at low costs. It is, of course, also clear that low ribbon production cost is not meaningful if solar cells of reasonable efficiencies cannot be manufactured from the material. Thus, the present status of knowledge about the relations between certain crystal growth variables and material properties, which is still rapidly changing, will be touched upon.

2 The Fundamentals of the EFG Method

2.1 General Principles

EFG can be viewed as one particular technique in a class of crystal growth processes that also includes, for instance, Czochralski and Stepanov growth, and which, generically speaking, may be called "surface tension controlled"[8]. More specifically, though, it has become known as a method which uses a "wetting die" for the shaping of the crystal. It is, therefore, instructive to consider at first the various steps of the process in a simple manner so that one may conceptually appreciate the various capillary and thermal conditions which must obtain to allow a crystal to grow. Hence, the first step is to insert a capillary die into a liquid which is kept near its melting point. On insertion, the liquid can rise to a total height which is given by the well-known capillary equation:

$$h_0 = 2\,\gamma \cos \phi_d / \rho g r. \tag{1}$$

Here, r denotes the radius of the capillary, or, for the case of two plates arranged side by side, the spacing between the plates. ρ is the density of the liquid, g the gravitational constant, ϕ_d the wetting angle of the liquid on the die material, and γ the surface tension of the liquid. However, to obtain edge defined crystal growth, the total equilibrium capillary rise cannot be utilized since firstly the time to achieve equilibrium rise tends to infinity according to:

$$\tau = \left[\frac{8\,\mu_v}{r^2 gg} h_0 \ell_n \left(\frac{h_0}{h_0 - h}\right) - h\right], \tag{2}$$

where μ_v is the dynamic viscosity; and secondly, the whole concept of edge definition depends on the liquid retaining sufficient pressure differential after rising to the end of the capillary, to still wet the capillary surface and spread to its outside edge when contacted with a seed.

We can now consider the first step in EFG growth to consist of establishing a liquid meniscus film at the top of a capillary – the die is now "filled", the liquid in the capillary being held above the melting point by proper adjustment of the crucible temperature. As a practical matter, then, it is quite apparent that the maximum allowed rise of the liquid over the level in the crucible, as well as the time needed to fill the capillary prior to the start of growth, depend on the die material chosen via the wetting angle and the capillary spacing, which is a variable design parameter.

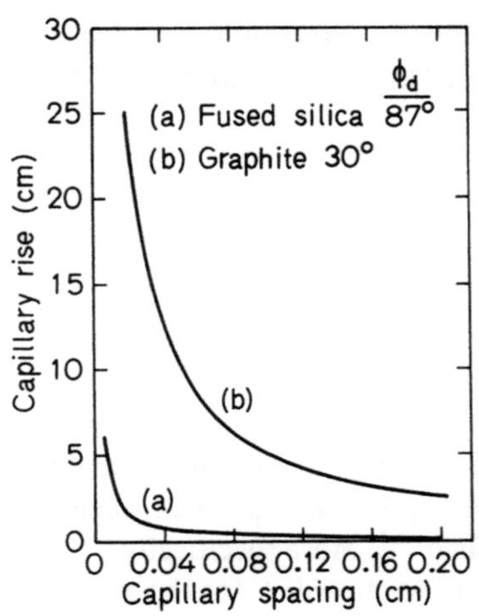

Fig. 1. Capillary rise (h_0) of liquid silicon as a function of capillary spacing (t_c or r) for graphite and quartz dies

Setting aside for the moment any ancillary considerations for the properties of the die materials, we may evaluate the overall capillary rise as a function of spacing for two die materials which can be used for silicon growth. The results obtained by using Eq. (1) for graphite ($\phi_d = 30°$) and quartz ($\phi_d = 87°$) are shown in Fig. 1. As can be seen from the figure, graphite appears to be quite practical as a die material from a capillary design standpoint, allowing the filling of dies of significant height using capillary spacings of 0.02 to 0.05 cm, whereas quartz presents definite problems, inasmuch as even a filling height of 5 cm would require spacings of the order of 0.01 cm. Also, Eq. (2) predicts very short filling times for graphite dies of such practical heights as 6 cm with a capillary spacing of ~0.02 cm, which can be reliably achieved. We may, therefore, conclude that only such die materials which are indeed "wetted quite well," i.e., have rather small wetting angles, are of practical value for EFG growth as it is generally practiced today. In fact, for the growth of silicon, carbon is the die material in universal use, and in spite of a widespread search for other candidates[9, 10] no better material which fulfills all requirements has yet been found. In the following discussions, we will thus assume that graphite (coated in situ with SiC) or other SiC-coated forms of carbon are the only practicable die materials available at this time.

The next step in crystal growth from such a capillary which is filled to the top with a liquid held above the freezing point would then be to contact the liquid surface with a solid seed. Since this seed conducts a finite amount of heat and re-radiates it to its surroundings, it is likely that a temperature adjustment is now required to keep the interface from freezing onto the capillary surface. Then, if the seed crystal which usually (but not necessarily) also consists of the material to be grown is wetted by the liquid, a meniscus is established which has its lower end anchored to the outer edge of the die and its upper end to the seed crystal (Fig. 2). Thus "edge definition" of the liquid meniscus is reached. If the seed crystal is then displaced upwards, the hydrostatic pressure in the meniscus film is reduced and a small volume of liquid flows upward through the capillary to equilibrate the pressure differential. For the crystal to continue growing, however, the die top temperature must now be lowered, i.e., a reduction in the amount of heat reaching the meniscus from the die is required to balance the heat of fusion released by the growing crystal which then moves downward at the expense of the liquid to restore the original interface position. Under

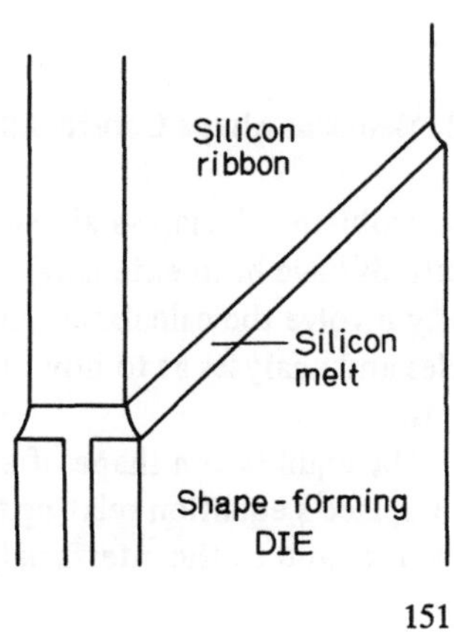

Fig. 2. Schematic of "Edge Defined" meniscus on a ribbon die

steady state conditions then, the solid can be pulled upward at a constant speed in such a way that the net rate of heat extraction (per unit area of interface) from the region of the liquid/crystal interface equals the rate of motion of the crystal times the latent heat of fusion per unit volume.

It is clear though, that now a momentary decrease in the rate of growth or any significant temperature fluctuation due to instabilities in the control system could upset this dynamic temperature balance.

However, EFG possesses an inherent stability against fluctuations in temperature or speed in the sense that the interface has a steady-state position for a range of temperatures and speeds. If the temperature of the liquid silicon at the die top is increased, the meniscus height increases; it reaches a new steady-state position so long as the meniscus is still within the range of capillary stability; if not, the solid ribbon pulls free and growth ceases. Conversely, if the temperature drops, the interface reaches a new steady-state position if the drop in temperature is not sufficient to reduce the meniscus height to zero at some point along the growth front, in which case the ribbon has a tendency to freeze to the die. Similarly, the interface moves in response to changes in speed. If these changes are sufficiently small, it settles down at a new position and growth is not interrupted. If they are large, the ribbon again pulls either free of the melt, or more commonly in practice, a "freeze" to the die occurs.

The operational question is, therefore, how sensitive is the position of the interface to fluctuations of temperature or speed; or does the interface remain within the limits which permit continuous growth when the temperature and the speed fluctuate within the range imposed by the thermal and mechanical design characteristics of the equipment? The basic criterion for growth stability, therefore, is the sensitivity of the meniscus height to fluctuations in temperature and speed. It can be shown that the crucial parameter which determines the sensitivity is the temperature gradient in the meniscus; if this is high, the sensitivity is low and relatively large fluctuations are permissible. Unfortunately, increasing the gradient in the liquid must be accompanied by decreasing speed of growth, if other conditions are unchanged.

It is then clear that the most fundamental problem in EFG growth is to establish the relationships among the various parameters involved (such as the thermal conditions at the die top, the solid-liquid interface, the meniscus shape, the crystal dimensions, and the die design) which all interact to allow the stable growth of crystals within a certain range of conditions.

2.2 Meniscus Shape Considerations

The problems of shape-stable growth in EFG and various other capillarity-governed methods have been extensively treated by Surek et al [8, 11, 12)]. These treatments generally involve the calculation of stable meniscus shapes from basic capillarity principles and analyses as to how these stable shapes react to perturbations of various kinds.

The equilibrium shape of a meniscus between the ribbon and die top is described by Laplace's equation relating the pressure difference ΔP across the meniscus to its curvature and to the interfacial tension γ:

$$\Delta P = \gamma\,(1/r_1 + 1/r_2). \tag{3}$$

Here, r_1 and r_2 are the principal radii of curvature of the meniscus and are defined to be positive when the center of curvature is inside the region of higher pressure. In a rectangular coordinate system where z measures the distance vertically downward from the crystal-melt interface and x measures distance horizontally from the center of the crystal (Fig. 3), the curvatures are given by:

$$\frac{1}{r_1} = \frac{d^2x/dz^2}{[1 + (dx/dz)^2]^{3/2}} \tag{4}$$

and

$$\frac{1}{r_2} = \frac{-1}{x[1 + (dx/dz)^2]^{1/2}}\,. \tag{5}$$

2.2.1 The Ribbon Geometry

For a ribbon geometry, calculation of the static meniscus shape from Eqs. (4) and (5) presents considerable difficulties. This is because, near the ends of the ribbon, the radius of curvature r_2 at $z = 0$ changes continuously from infinity along the sides of the ribbon to approximately $-t/2$ at the very end of the ribbon (t is the ribbon thickness). A mathematical solution for the complete meniscus shape has, therefore, never been attempted[11)], but, instead, the two separate problems of the meniscus against a vertical flat surface and that against a vertical cylinder of radius t/2 have been considered. These solutions are taken to be representative of the meniscus shapes at the side and the very ends of the ribbon, respectively.

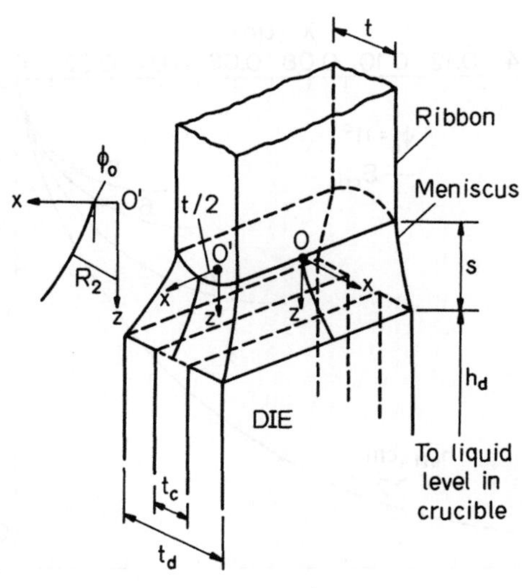

Fig. 3. Definition of meniscus shape variables which are used in calculations

The hydrostatic pressure difference at any point (x,z) on the meniscus is given by:

$$\Delta P = \rho g\,(h_d + s - z) + P_V, \tag{6}$$

where ρ denotes the density difference between the media on either side of the meniscus, g is the gravitational constant, h_d is the height of the top of the die above the melt level in the crucible, s is the vertical distance (or meniscus height) between the top of the die and the bottom of the ribbon, and P_V represents the pressure drop associated with viscous flow in the capillary and in the meniscus film. Since s, z, and P_V are generally so small as to be negligible, the terms have been combined into a parameter H_{eff}, i.e., the effective height of the growth interface above the liquid level in the crucible, viz.,

$$\Delta P \approx \rho g\, H_{eff}. \tag{7}$$

Thus, H_{eff} is inversely proportional to the total hydrostatic head in the meniscus.

Equations (4), (5), (6), and (7) may be combined and integrated by numerical techniques to determine the meniscus shape function x(z). The boundary conditions for the two problems are x = 1/2 and dx/dz = tan ϕ_0 at z = 0. In addition, r_2 is negative for the simulated ribbon end problem and infinity for the ribbon side. The density (ρ = 2.53 g/cm^3), surface tension (ρ = 720 erg/cm^2), and equilibrium wetting angle of liquid to solid silicon (ϕ_0 = 11°)[13] are materials parameters.

On the ribbon side, then, the meniscus shape can be calculated as a function of h_{eff}, as shown in Fig. 4. If one assumes a thickness (t) of 0.02 cm, which is somewhat typical for present day EFG silicon ribbon, calculation of the end meniscus shapes leads to the results shown in Fig. 5. One notes that the radius of curvature in the vertical plane is always much smaller in the case of the end meniscus than that of the side meniscus for the same value of H_{eff}.

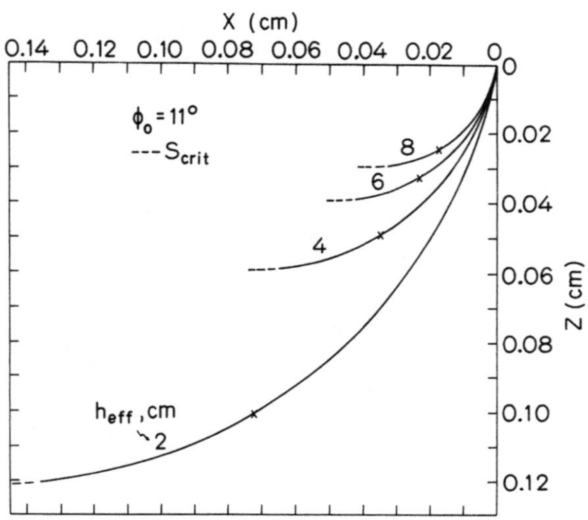

Fig. 4. Calculated meniscus shapes at the ribbon side as a function of H_{eff}

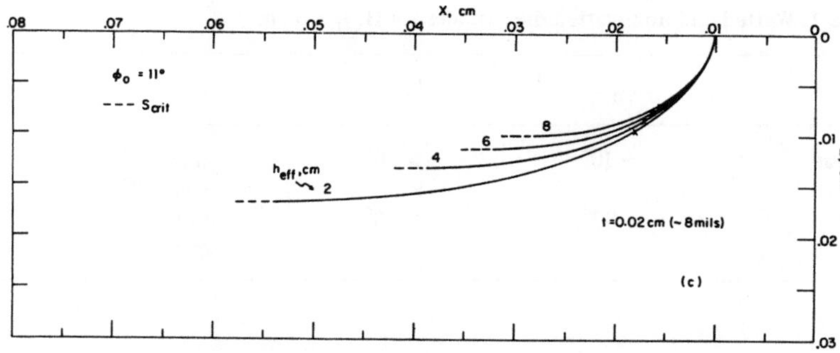

Fig. 5. Calculated meniscus shapes at the end of a 0.02 cm thick ribbon as a function of H_{eff}

Next, the attachment of the lower end of the meniscus to the die is considered. Figure 6 defines the angles θ_w and θ_{nw} which distinguish the wetted and non-wetted surfaces of the die. The angle ξ designates the inclination of the meniscus relative to the horizontal, and θ_d designates the contact angle of the liquid on the die material. To calculate the maximum surface inclination for wetting θ_w, we first find ξ for the steady-state configuration using the approximation that the meniscus is an arc of constant radius subtending an angle η. From the figure, it is clear that $\xi = 90^\circ - \eta - \phi_0$. For this surface to be wetted, we must have $\xi - \theta_w > \theta_d$; therefore,

$$\theta_w < 90^\circ - \eta - \theta_0 - \theta_d. \quad (8)$$

To calculate the minimum inclination for non-wetting θ_{nw}, we assume the worst possible spreading situation; i.e., $\xi = 90^\circ$. In this case, we have

$$\theta_{nw} > 90^\circ - \theta_d. \quad (9)$$

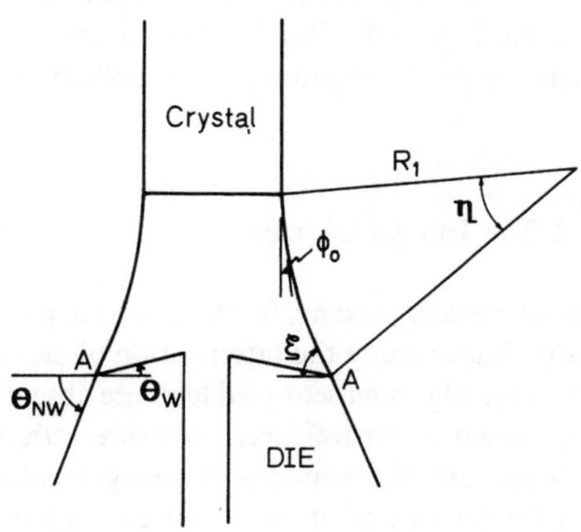

Fig. 6. Geometric relationships of a ribbon in contact with a die indicating quantities which are useful in the calculation of wetting and non-wetting die shapes

Table 1. Wetted and non-wetted die surfaces for $H_{eff} = 6$ cm

Die	ϕ_d	θ_{nw}	η	θ_w
Carbon	~30°	> 60°	sides 43°	< 6°
			ends 33°	< 16°
Silica	~87°	> 3°	sides 43°	< −57°
			ends 33°	< −41°

Table 1 gives the value of θ_w and θ_{nw} for the two potential die materials considered before and ϕ_0 of 11°. Here, we consider a die with $H_{eff} \sim 6$ cm, and meniscus heights of ~0.03 cm along the ribbon sides and ~0.01 cm at the ends of a 0.02 cm thick ribbon. Negative values of θ_w signify surfaces which slant toward the center of the die. There are also certain die designs (such as the "knife-edge die") in which θ_w is no longer defined. However, for most practical designs in use today, θ_w is in fact 0°.

The shape of the die in its width direction should also be considered (cf. "end" values in Table 1) as the meniscus shape calculations show the maximum meniscus height at the ribbon ends is always less than that at the sides of the ribbon. Since for a flat die, in a first approximation, the freezing isotherm is considered to be parallel to the die top, it is quite apparent that now the available meniscus height all across the die is limited by the small value required by the ribbon ends. Any slight fluctuations in the ribbon pull rate or the thermal environment may, therefore, displace the freezing isotherm out of the range of stable meniscus heights at the ribbon edges.

To counteract this, the thermal field around the edges has to be properly controlled, and one way of achieving such control is by shaping of the die top contours. This problem has been considered in detail theoretically[8)] and has also led to the evolution of fairly complex die shapes[14, 15)] which demonstrably aid stable growth of ribbon under specific growth conditions, and which may also be used to influence more subtle ribbon properties, as shown by Ciszek in this volume.

However, the complications introduced into ribbon growth by the "end problem" make a ribbon a poor model for quantitative experiments in testing the theories of shape-stable growth. This difficulty is heightened by the fact that it is difficult to match a thermal environment to the ribbon shape.

2.2.2 The Tubular Geometry

A tubular shape presents fewer of the difficulties just mentioned and is, therefore, nearly ideal to study the interrelation of certain growth variables[16)]. Firstly and most obviously, edge meniscus problems are absent. Also, a circularly symmetrical thermal environment is relatively easy to create with induction heating equipment.

Therefore, the meniscus shape arguments considered above are redefined specifically for the case of tubes so that a basis is created to treat the various fundamental

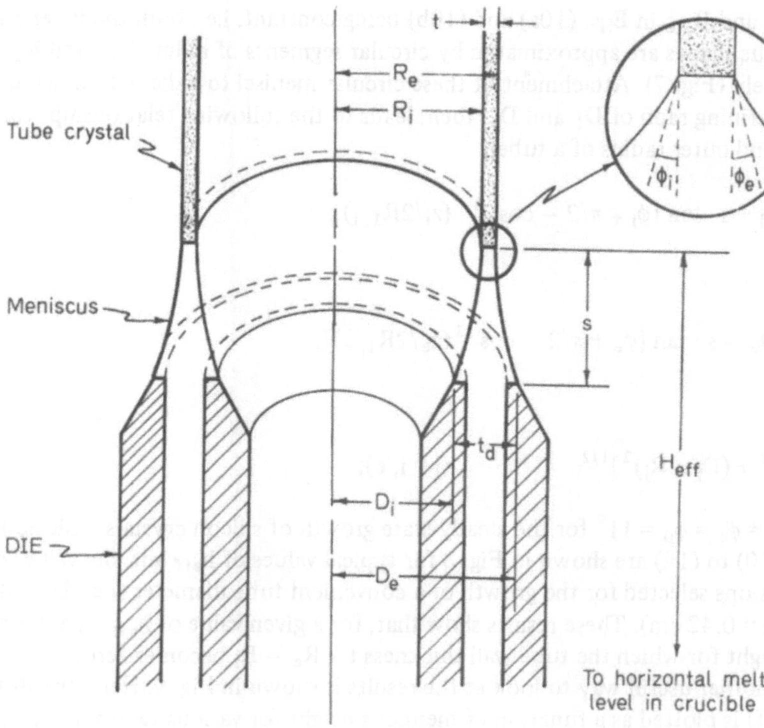

Fig. 7. Redefinition of meniscus shape variables for the case of tube growth

relations in EFG growth experimentally using a simple tubular shape[17]. With the aid of Fig. 7 then, we can rewrite Eqs. (4) and (5) for tubes in a simplified manner as:

$$R_{2,e} = -R_e/\cos\phi_e \tag{10a}$$

and

$$R_{2,i} = R_i/\cos\phi_i, \tag{10b}$$

where the subscripts "e" and "i" are used for the external and internal variables in the tube growth system. R_e and R_i are then the outer and inner radii, respectively, of the tube crystal. For stable growth then, where R_e, R_i and the tube wall thickness ($t = R_e - R_i$) are all constant, the meniscus angles ϕ_e and ϕ_i, measured between the applicable meniscus and the crystal growth axis at the growth interface in Fig. 7, must both be equal to the characteristic angle ϕ_0 of 11° discussed above.

A useful approximation is to set the R_2 terms given by Eqs. (10a) and (10b) constant over the entire meniscus. This is consistent with the assumption that Δp = constant over the entire meniscus too. Combining the results of these assumptions leads to $R_{2,i}$ and $R_{2,e}$ in Eqs. (10a) and (10b) being constant, i.e., both the inner and outer meniscus shapes are approximated by circular segments of radius $R_{1,i}$ and $R_{1,e}$, respectively (Fig. 7). Attachment of these circular menisci to a die with inner and outer edge-defining radii of D_1 and D_e, then, leads to the following relationships for the inner and outer radius of a tube:

$$R_i = D_i + s \cdot \tan\left[\phi_i + \pi/2 - \cos^{-1}\ (z_i/2R_{1,\,i})\right], \tag{11a}$$

and

$$R_e = D_e - s \cdot \tan\left[\phi_e + \pi/2 - \cos^{-1}(z_e/2R_{1,\,e})\right], \tag{11b}$$

where

$$z_j = [s^2 + (D_j = R_j)^2]^{1/2} \qquad (j = i, e); \tag{12}$$

and $\phi_i = \phi_e = \phi_0 = 11°$ for the steady-state growth of silicon crystals. Solutions of Eqs. (10) to (12) are shown in Fig. 8 for typical values of h_{eff} and the values of die dimensions selected for the growth of a convenient tube diameter (i.e., $D_e = 0.48$ cm and $D_i = 0.42$ cm). These results show that, for a given value of h_{eff}, there is a meniscus height for which the tube wall thickness $t = R_e - R_i$ becomes zero.

Another useful way to look at the results is shown in Fig. 9. Here, the difference $(t_D - t)$ is plotted as a function of meniscus height for various values of H_{eff}; the tube wall thickness increases from right to left in the figure. The quantity $t_d = D_e - D_i$ represents the edge-to-edge thickness of the die top (see Fig. 7). The above

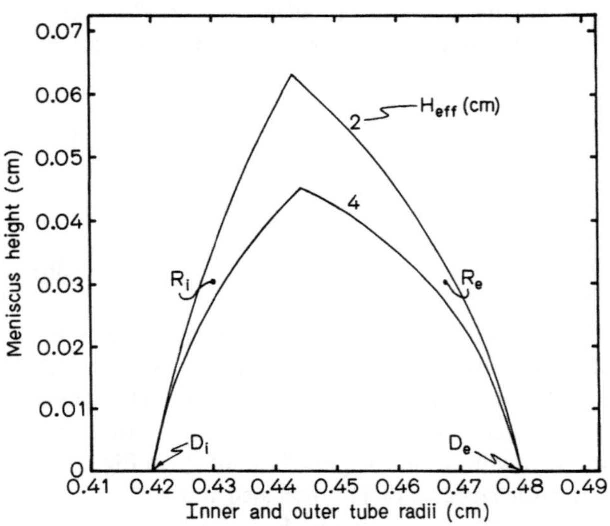

Fig. 8. Calculated values for the variation of inner and outer tube radii (R_i and R_e) with the meniscus height. The two values of H_{eff} are typical for small diameter tube growth

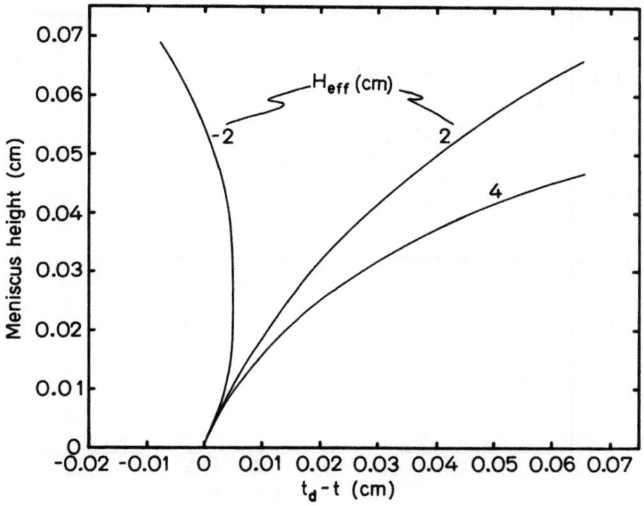

Fig. 9. Calculated relationships between meniscus height and tube wall thickness as a function of H_{eff} in silicon tube growth from graphite dies

analysis can then be used to show that the inner and outer dimensions of the tube (and hence the wall thickness) are uniquely determined by the values of the meniscus variables D_e, D_i, ϕ_0, H_{eff}, and s. The angle ϕ_0 is a fixed materials parameter; the choice of die dimensions is an experimental variable which is influenced by considerations such as the chosen tube dimensions, capillary rise, etc. Basically, D_e and D_i (and hence t_d) are assumed to remain constant during a given growth run, i.e., effects of chemical reactions between the melt and die are negligible. The remaining variables (H_{eff} and s) are both realtime growth variables. The initial value of H_{eff} is determined by the die height, the crucible and die geometries, and the silicon charge volume. During growth, H_{eff} increases as the melt level in the crucible decreases. Thus, in experiments where the meniscus height is kept constant, it is expected that the wall thickness would decrease as longer segments of tube are grown. Figure 10 indeed shows the excellent agreement that was found between the results predicted from Eqs. (11) and (12) and the experiments[17].

Another important result which is obtained from these meniscus shape calculations is that the tube-shape is inherently stable in the EFG process. A quantitative measure of the stability of the crystal dimensions is given by the derivatives $|dR_d/ds|$ and $|dR_i/ds|$ which can be obtained from Fig. 8, or the quantity $|dt/ds|$ from Fig. 9. Clearly, lower absolute values of these slopes imply a lesser dependence of the tube dimensions on meniscus height, and hence better control of the dimensions. Stability is then seen to increase as s or H_{eff} decreases, i.e. it increases with an increasing hydrostatic pressure inside the meniscus.

Another useful conclusion that can be drawn then is, that the design of growth apparatus should emphasize low values for H_{eff} in order to maximize growth stability over a wide range of meniscus heights.

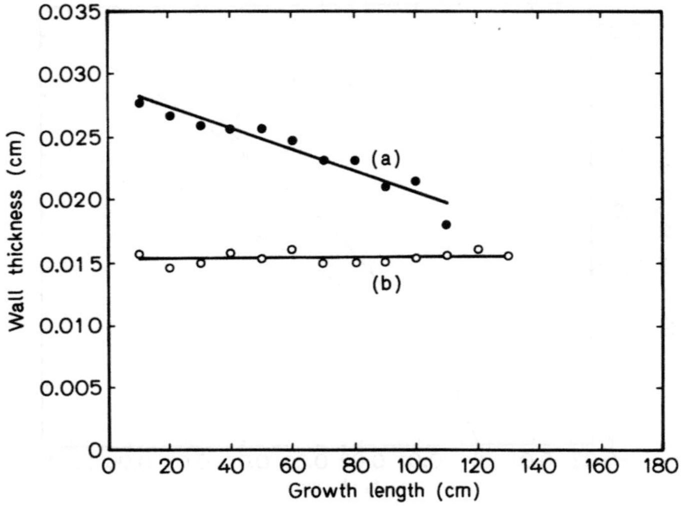

Fig. 10. Experimentally determined variation of tube wall thickness as a function of H_{eff} in the growth of silicon tubes from graphite dies, **a** shows the result of an experiment where the meniscus height was kept constant and thus H_{eff} increased in the course of the run, due to depletion of melt in the crucible. **b** shows, how through proper adjustment of meniscus height during the run, according to theory, the wall thickness can be kept constant

2.3 Thermal Considerations

In crystal growth, the macroscopic and microscopic quality of the resulting product is in general strongly influenced by the shape of the solid-liquid interface, which must therefore be closely controlled. The volume output on the other hand depends on the efficacy with which the heat of crystallization is removed from that interface. The temperature gradient in the liquid has important consequences as well in determining growth stability in a geometrical sense and furthermore affects the materials quality in that it controls constitutional supercooling and related effects. Also, the temperature gradients in the solid crystal are of concern, since they may cause stresses to occur, which can even lead to fracture. Also, the cooling environment of a crystal may cause chemical changes in the solid, which often cannot be ignored if the details of the electronic properties of the resulting crystals are of concern.

By calculating a three-dimensional dynamic heat balance during the crystal growth process, it is possible in principle to specify the overall thermal behavior of the crystal. Since for such an undertaking, it would be necessary to consider all thermal inputs and outputs, one would also gain important information on thermal variables which potentially could be used for process control.

Such a heat balance calculation then requires knowledge of all possible conductive and radiative contributions to the heat input at the interface, not only from the liquid which conducts heat and also carries heat to the meniscus and the interface, but also from the die and the various other shunt heat paths of the growth setup. All these factors, many of which depend very strongly on the details of the apparatus, have to be considered, along with the heat of fusion evolved, for the input side of the balance equation. Similarly, all heat removal mechanisms must be known quantitatively, and a particularly difficult term to evaluate is the convective transport caused by the inert gas present in the furnace. Also, there is a question as to how much heat is removed at the interface by conduction and how much by radiation (light piping!). This is important not only because these various heat removal mechanisms have somewhat different efficiencies, but also because the radiative loss by light piping has, on one hand, the beneficial effect that it does not require a temperature gradient in the solid close to the solid-liquid interface for the removal of heat and thus would reduce one cause of stress in this region of the crystal. On the other hand, the piped heat loss is a fixed quantity which cannot be manipulated by control of the thermal environment normal to the growth axis, which may include forced cooling using gas jets as well as the creation of low temperature radiation environments.

These types of questions are still being analyzed since they are of particular significance to high speed growth of ribbon, and preliminary consideration had been given to the light piping terms fairly early on[18]. Also, under certain particular conditions, the system can be analyzed in a useful way by considering the die as a heat flow limiting thermal resistor[19]. In the main, however, these problems are still awaiting solutions and are considered only qualitatively and intuitively during equipment construction and growth experiments which, in the following discussions, are still analyzed by using simplified models which do not consider factors such as light piping, etc.

For all these reasons, then, it is obvious that a three-dimensional thermal balance calculation is not only quite difficult to perform, but would probably be of no practical use because it is likely to be extremely dependent on the details of the equipment under consideration. However, it has been shown[10] that even a one-dimensional thermal balance analysis can give important insights because it can provide some feeling at least for the upper limits on the achievable growth velocity, and it also allows some estimate of the shape stability which, as we have seen, depends on the thermal balance via the meniscus height (s) when H_{eff} is constant.

However, as noted before, the situation at the ribbon edge is complicated from a capillarity standpoint, which is compounded by the fact that, in addition, at the ribbon edge the thermal environment is also hard to specify. Hence, recourse to tube growth for an experimental test of the theoretical predictions again seems useful.

2.3.1 Tube Growth

In this section then, we will compare the results predicted by the simplified one-dimensional heat flow analysis of Swartz et al.[10] with experimental results obtained during the growth of small, thin-walled tubes. The experiments have been described also in the recent paper by Eriss et al[17].

They derived for tubes the following expressions relating the wall thickness to the growth rate and the gradient in the meniscus:

$$LV_g = \alpha f(\beta)\, t^{-1/2} - k_\ell G \tag{13}$$

with

$$\alpha = (\sigma \epsilon k_m T_m{}^5)^{1/2} \text{ and } \beta = T_m/T_0. \tag{14}$$

Finally, the normalization function f (β) can be written as:

$$f(\beta) = [1 - \beta^{-4}(1 + 4\ell n\beta)]^{1/2}. \tag{15}$$

The terms are used as follows:

L = Latent heat of fusion per unit volume of crystal.
V_g = Growth rate.
k_m = Thermal conductivity of the solid crystal at the melting point, T_m.
T_0 = Effective temperature of the radiating environment.
k_ℓ = Thermal conductivity of the liquid.
σ = Stefan-Boltzmann constant.
ϵ = Emissivity of the tube surface.
G = Gradient in the liquid at the growth interface.

The latter is a function of the meniscus height and of the difference between T_m and the die top temperature T_D, and it can be approximated as

$$G = \left(\frac{T_D - T_m}{s}\right)\left(\frac{t_D/t - 1}{\ell n(t_D/t)}\right). \tag{16}$$

A value for the meniscus height s may then be chosen by specifying V_g and T_D. The latter, in a simple system, can be most easily done by adjusting the crucible temperature.

It is clear then that there exists no unique relationship between the growth rate and the tube wall thickness, the latter of which is given only by the value of the meniscus height and the shape variables from Sect. 2.2.2.

Thus, the same wall thickness can be grown at rates ranging from zero (where the gradient in the liquid is maximum) to a rate

$$V_{g,\max} = \alpha f(\beta)\, t^{-1/2}/L, \tag{17}$$

where the gradient in the liquid is zero.

Since in tube the growth rate is not limited by edge stability considerations (as in ribbons), these relationships can now be experimentally tested. Figure 11 (based on Ref. 17) shows a plot of the maximum growth speeds obtained at various thicknesses, along with two curves calculated using the relations discussed above for two different radiating environments. It can be seen that the experimental points agree extremely well when an environment at 1250° K is assumed.

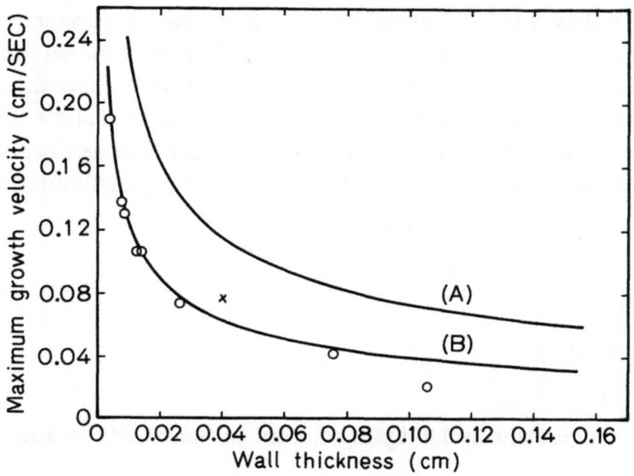

Fig. 11. Experimentally determined maximum growth velocity of a tube as a function of wall thickness. The solid lines were obtained from theory using for **(A)** a radiative environment temperature (T_0) of 0 °K, and for **(B)** one of 1250 °K. The latter correlates very well with the experiments. Also shown (point x) is a result for a ribbon growth experiment where the thickness was determined as 0.04 cm and the radiative environment temperature as 1100 °K. (See text)

These experiments thus confirm quantitatively the relations between the capillary variables and the simple one-dimensional heat flow model. However, they also highlight the problem which we touched upon in the general discussion of Sect. 2.3, i.e., the tube crystal growth system as designed, arbitrarily behaved like a black body at 1250° K (at least at the point where the wall thickness was observed). Hence, if higher growth speeds were to be realized, an equipment design which achieves a much lower environmental temperature to receive radiation from the ribbon would be needed. A closer reading of the careful observations of these authors with respect to circumferential wall thickness variations makes it even more plain that the details of the equipment construction play a major role as controlling parameters in EFG crystal growth.

2.3.2 Growth of Wide Ribbon at High Rates

For the growth of ribbons at rates similar to those of tubes, in principle the same relations apply, except that, to achieve the most stable conditions for growth, one generally wishes to utilize the maximum edge meniscus heights allowed by capillarity. When using a flat die of uniform cross section, however, that generally requires adjustments in the thermal fields at the die top such that the center meniscus is much higher and the ribbon thickness in the center accordingly less than at the edges. Still, in principle, a point-by-point heat balance using a one-dimensional model could still be used. At the edges, certain complications arise, though, because the edge has a greater surface, per unit cross-sectional area, than the center of the ribbon.

Kalejs et al.[20] showed, by using thickness measurements on ribbons grown at various speeds, that the one-dimensional heat flow model can still give valuable information. In a modified form, it can also serve to estimate the die top temperature along the ribbon face as well as at the edge and then allow the calculation of certain growth parameters using the measured ribbon thickness profiles. The difference in temperature between the die top (T_D) and the growth interface (T_m) is expressed as:

$$T_D - T_m = (s'/k_L)\left[\frac{\alpha f(\beta)}{t^{1/2}} - \rho VL\right],$$

where (18)

$$s' = \frac{t\ell n\,(t_D/t)}{(t_D - t)}\,.\,s$$

and the gradient in the liquid adjacent to the growth interface becomes

$$G = (T_D - T_m)/s'. \quad (19)$$

s' is related to s by a geometric factor which can take into account variations in temperature gradient in the meniscus due to changes in its cross sections which arise from thickness differences between the ribbon t and the die top t_D at each point along the die.

The factors α and $f(\beta)$ have been discussed previously in Sect. 2.3.1 when tube growth was considered, and, in the present case, the factor β again contains T_0, the effective temperature of the radiating environment. However, here T_0 becomes more difficult to specify. Because of the geometry of the ribbon and the resulting complexity of the thermal environment in which high speed ribbon growth experiments take place, only an overall average of T_0 can be calculated from the maximum growth speed capabilities determined for a system, which here lead to a value of 1100 °K which gives $f(\beta) = 0.7$. Average ribbon thicknesses for both the center and the edge

Table 2. Die top temperature differences calculated from one-dimensional heat transfer model (Eqs. (18) and (19) in Text). $T_0 = 1100$ °K, $\epsilon = 0.54$, $k_L = 0.6 \times 10^7$ ergs/cm^2 – °K, $\rho L = 4.14 \times 10^9$ ergs/cm^3, and $t_D = 0.066$ cm

Growth speed (cm/min)	Average thickness measurement location	Average thickness (cm)	Meniscus height (cm)	G (°C/cm)	$T_D - T_M$ (°C)
3.4	ribbon face	0.033	0.036	220	5.6
3.4	ribbon edge	0.056	0.018[a]	80	1.3
3.8	ribbon face	0.026	0.040	255	6.2
3.8	ribbon edge	0.044	0.020[a]	95	1.5
4.7	ribbon face	0.020	0.043	248	5.5
4.7	ribbon edge	0.040	0.022[a]	17	0.3

[a] Estimated average meniscus height for a region within 0.5 cm of the ribbon edge

of ribbons grown at different speeds can now be calculated, and an example is found in Table 2.

If one now solves Eq. (18) for the ribbon thickness which can support a maximum growth speed of 4.7 cm/min with T_0 = 1100 °K, the value of 0.043 cm is found. This is very close to the ribbon edge value of 0.040 cm which is measured and therefore suggests that the speed capabilities of the system are limited by the ribbon edge, and a comparison with the results on tubes, presented in Fig. 11 (Sect. 2.3.1), reveals very satisfactory agreement as well. This close agreement between the measured and the calculated values for these cases supports the validity of using the simple one-dimensional heat transfer approach to further the understanding of the system during growth.

However, there are important aspects of dynamic ribbon edge stability during growth which appear to be entirely beyond the scope of the one-dimensional heat flow model and for which at present no satisfactory theory exists. Some of these "dynamic" phenomena will now be discussed, and the current means by which they have been overcome will be described. All these experiments were performed using the die shown in Fig. 12 in a system which provides a means for the separate addition of heat at both ends and in the center through the use of two resistance heated end heaters and a face heater for the die (Fig. 13). For instance, when a ribbon of 7.4 cm width, growing at a steady-state velocity of 3.4 cm/min is subjected to a sudden increase in pull speed up to 3.8 cm/min, not only the meniscus height is raised, but the ribbon starts to taper to a width of 7.2 cm at which it is grown further in steady state. To restore growth at a constant width of 7.4 cm at this speed now requires adjustments to the power level of the end heaters. A further increase in speed from 3.8 to 4.7 cm/min results in a decrease in ribbon width from 7.4 cm to 6.8 cm over ~8 cm of growth. When the end heaters are again adjusted, constant width growth up to 7.1 cm can be restored.

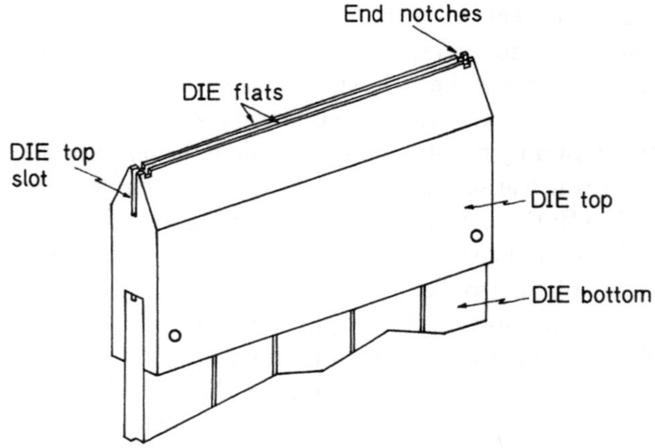

Fig. 12. Schematic of a "two-piece" die used in the growth of 7.5 cm wide ribbons. The upper piece is machined from a rectangular piece of graphite into which a 0.02–0.05 cm wide capillary is machined. The die flats are typically 0.05–0.01 cm. The bottom contains saw cut grooves, machined into a graphite slab by a simple saw cut. The two pieces are pinned together with graphite pins at (o). The end notches serve to stabilize the ribbon edges during growth

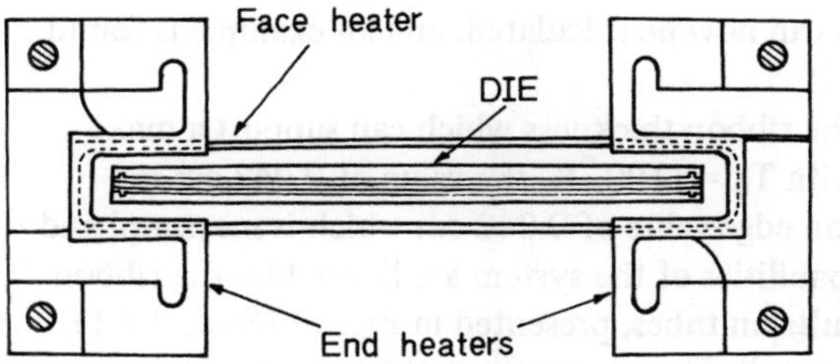

Fig. 13. Top view of the die (Fig. 12) surrounded by "face" and "end" heaters

The mechanisms operative during responses of the ribbon edge to speed changes are not understood. However, it is believed that the temperature gradients at the ribbon edges are of importance and, for the particular die design and face and end heater arrangements shown in Figs. 12 and 13, (more discussion is provided in Ref. 20). Even then, an empirical demonstration of ribbon edge position control and stability conferred on the growth process as a result of such a heater configuration is of great importance for the continuous growth of ribbon. It was also recognized quite early that it could become the basis for an automated system to maintain constant width growth over long periods of time without operator control[21].

2.3.2.1 Principles of Residual Stress Generation in Ribbon

Residual stresses in EFG grown ribbon originate from the manner in which the ribbon is cooled from the melting point to room temperature[20, 22]. During cooling, each element of the ribbon will contract proportionally to its own temperature fall. If adjacent elements contract differently, they will restrain each other's contraction; hence, thermal stresses result. By considering the thermal contraction of adjacent longitudinal elements of the ribbon, it can be readily shown that a stressfree ribbon would remain stress-free if it is cooled in a uniform temperature gradient (assuming, of course, that the thermal expansion coefficient is independent of temperature). If the temperature gradient is nonuniform (i.e., non-linear), adjacent elements can be made to fit only by lateral deformation in the plane of the ribbon. Thus, the thermal stresses which arise will be proportional to the curvature or bending that must be imposed to fit the elements together, if no in situ thermal stress relief were to occur.

Theoretical analysis of thermal stresses in ribbon growth[20, 22] considers the case of a thin ribbon (so that a plane stress solution is applicable) with arbitrary temperature distributions along the length and across the width of the ribbon. The solution is obtained either in analytic form (for simple forms of the temperature distribution) or by numerical analysis on a computer. For the usual case of interest, where the temperature profile is assumed to be a function of the vertical coordinate x only, the variation of the stress components σ_{xy}, σ_{yy}, and σ_{xx} with the variable y along the ribbon width is shown in Fig. 14. The normal stress component in the longitudinal direction (σ_{xx}) is seen to lead to a compressive stress in the outer portion of the ribbon, whereas the center of the ribbon is in tension. The magnitudes of the stress components as a function of the vertical coordiante depend on the functional form T (x)

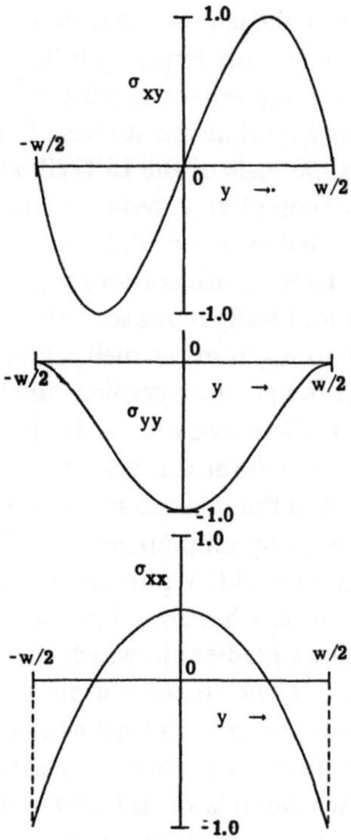

Fig. 14. Variation of stress components σ_{xy}, σ_{yy}, and σ_{xx} with the coordinate y across the ribbon width w

of the temperature profile. For T(x), appropriate for radiation of heat from the ribbon surface into a 300 K environment, calculated thermal stresses are found to be far in excess of the yield stress of silicon at the temperatures under consideration. It is, therefore, not surprising that excessive residual stresses are encountered in ribbons which are grown under these conditions. The theory further predicts that the thermal stress components should be proportional to $(\text{width})^n$, where $n \simeq 2$ to 4. It is important to note here, however, that this conclusion has to be modified significantly if one considers the possibility of transverse thermal gradients in the ribbon.

Such thermal stress calculations describe the forces that would tend to cause plastic deformation and the dependence of these forces on thermal profiles and on ribbon geometry. If plastic deformation did not occur during ribbon growth, then the above theory would be immediately useful as it predicts that either the ribbon breaks in a region where the thermal stress is too high, or the ribbon emerges at room temperature with no residual stress. Since, however, plastic deformation does occur, the calculation of the residual elastic stress in the grown ribbon requires solution of the complete elastic-plastic problem, which involves the inclusion of time-dependent plasticity effects into the theory.

Plastic deformation acts to relieve the thermal stress during growth; this relief may be complete or partial depending on time at temperature. Thus, residual elastic strain in the ribbon can be avoided if there is either complete relief of stress or no relief of stress (i.e., no plastic strain) during the whole of the change of temperature of each element of the ribbon from the melting point to room temperature. We are, therefore, concerned with the minimum time required, at various temperatures, to allow complete stress relief, or the maximum time during which no stress relief occurs. From experimental studies of stress relaxation in EFG ribbons, it is expected that, on the time scale of the EFG ribbon growth, stress relief in the ribbon would be effectively complete at temperatures above 1200 °C, and that almost no stress relief would occur below about 700 °C.

These considerations suggest that the vertical temperature profile in the ribbon should, ideally, consist of three zones to permit fast growth of stress-free ribbon. The first zone (from the melting temperature to ~1200 °C) should consist of an initially large temperature gradient to allow high rates of growth and a very rapid decrease of the gradient away from the growth interface. Plastic strain in this zone should occur fast enough for the stress to be completely relieved. Of concern in this temperature region is the minimization of plastic strain-induced imperfections. This can be achieved by minimizing second and higher order derivatives of the vertical profile, or perhaps could be possible by countering the effects of the vertical profile with an appropriate horizontal profile. The second zone should be a region of constant temperature gradient in which the stress-free ribbon cools from ~1200 °C to below ~600 °C; no stress or strain is introduced in the ribbon in this zone. The third zone should consist of a further decrease in gradient that is needed to reach zero gradient at room temperature. No plastic deformation can occur in this latter zone; therefore, if the ribbon is stress-free when it enters it, it will again be stress-free when it leaves this zone at room temperature. The crucial question in this third zone is the elastic (thermal) stress which is developed; if this stress does not fracture the ribbon, it will then disappear as the gradient becomes zero at room temperature.

2.3.2.2 Approaches to Stress Management During Ribbon Growth

As discussed above, analytical expressions have been derived[10)] for the vertical temperature profile in the ribbon for the case where the radiating environment seen by the ribbon is at a uniform temperature T_0. The analysis has been useful in relating the ribbon thickness to the growth rate and the gradient in the liquid ahead of the growth interface; the effective environment temperature has been determined from correlations of experiment and theory[20)].

Other than describing the gradient in the ribbon at the interface, however, this analytic approach is insufficient to describe the vertical profile in the ribbon to the extent that useful determinations of the thermal stresses can be made. The main problems lie in the oversimplification of the radiating environment and the assumption of a stationary ribbon (i.e., neglecting possible light piping, for instance, as well as the specific heat carried by the moving ribbon) in the calculation. The inclusion of a realistic thermal environment (which can also be readily correlated to the experimental crystal growth setup) and of more specific heat transport mechanisms into the calculations can best be handled by computerized numerical analysis. The program which was developed has been used to study the feasibility of the three-zone concept

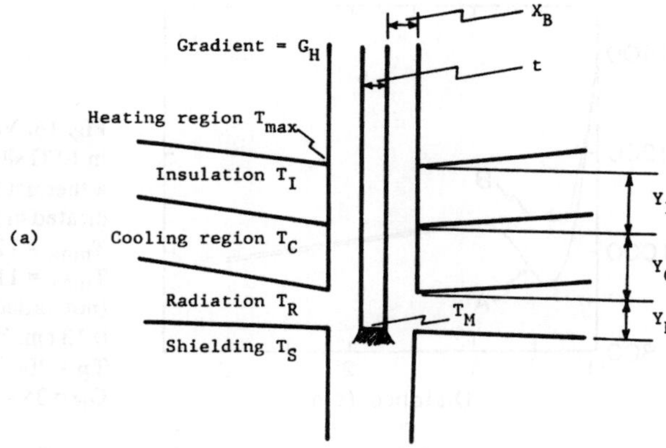

Fig. 15. Schematic of a cooling-heating arrangement used to prevent ribbon stress

for the vertical temperature profile described earlier in this section, and to design the necessary cooling/heating environment to achieve the desired profile.

Figure 15 shows schematically the model of the thermal environment of the ribbon. As indicated in the figure, the thermal environment consists of five distinct regeions; this is representative of the experimental arrangement ("growth system cartridge") described later on. The temperatures in the shielding and radiation regions (T_S and T_R, respectively) correspond to the effective temperatures in the regions beneath the growth interface and in the low angle viewing region beneath a cold block (or cooling region in the figure). The actively powered afterheater region is designed to give a linear gradient G_H along the heater block; the power is adjusted to give a maximum temperature T_{max}, as shown in Fig. 15. The cooling region is assumed to be at a constant temperature T_C; the temperature of the insulation regions is assumed to be $T_I = 1/2\,(T_C + T_{max})$. The geometrical variables in the model (Y_B, Y_C, T_I, and X_B) are all defined in the figure; "t" represents the ribbon thickness.

The computer model used to calculate the steady-state temperature profile in the ribbon for a given set of conditions in Fig. 15 considers conductive and convective heat transport in the ribbon, and radiative and convective cooling at the ribbon surface. Heat flow in the ribbon is assumed to be one-dimensional (also, ribbon width $\gg$ ribbon thickness). The radiative heat flux is determined from the temperatures, emissivities and angle factors of the surfaces, which comprise the thermal environment of the ribbon. The convective cooling (or heating) of the ribbon surface involves a heat exchange through the gas ambient between the ribbon and the cooling/afterheating blocks at each point.

Figure 16 shows the results for two sets of values of the parameters (listed in the figure caption). Curve A is a typical profile for cases where extensive cooling of the ribbon takes place in the radiation, cooling and insulation regions before the ribbon is reheated by the afterheater. Within the afterheater region, the temperature profile in the ribbon is essentially the same (i.e., linear) as in the afterheater block. The "undercooling" of the ribbon is a very sensitive function of the extent (Y_C) and the position (Y_B) of the cooling block, as well as its temperature and distance from the ribbon surface. Desired decreases in the values of Y_B, Y_C, and Y_I may be limited, of

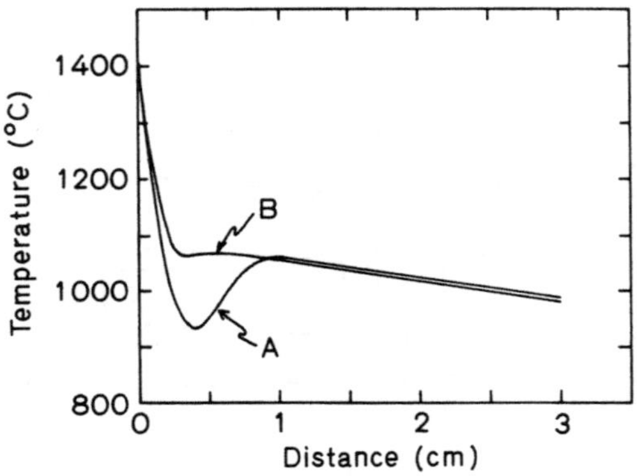

Fig. 16. Vertical temperature profiles in EFG silicon ribbon calculated from a thermal model using variables as indicated in Fig. 15. **(A)**: $Y_C = 0.38$ cm, $T_{max} = 1200$ °C; **(B)**: $Y_C = 0.13$ cm, $T_{max} = 1100$ °C. Other parameters (not varied): $Y_B = 0.1$ cm, $Y_I = 0.13$ cm, $X_B = 0.075$ cm, $T_S = 1200$ °C, $T_R = 700$ °C, $T_C = 200$ °C, $T_I = 700$ °C, $G_H = 35.4$ °C/cm

course, by practical design considerations. Curve B in Fig. 16 shows the temperature profile for a smaller value of Y_C; the maximum temperature in the afterheater is also lower than for the previous case. This latter profile closely approximates the ideal three-zone profile described in the previous section (only the first zone and a portion of the second zone are shown in the Figure). Both thermal configurations in Fig. 16 would permit stressfree growth of 0.025 cm thick ribbons at speeds of about 7.5 cm/min.

3 Growth Equipment

In its most simple form, EFG growth equipment is perhaps represented in the growth of small diameter thin-walled tubes[16)]. The parts are shown schematically in Fig. 17 in which a growth setup is placed inside an induction-heated double-walled quartz chamber.

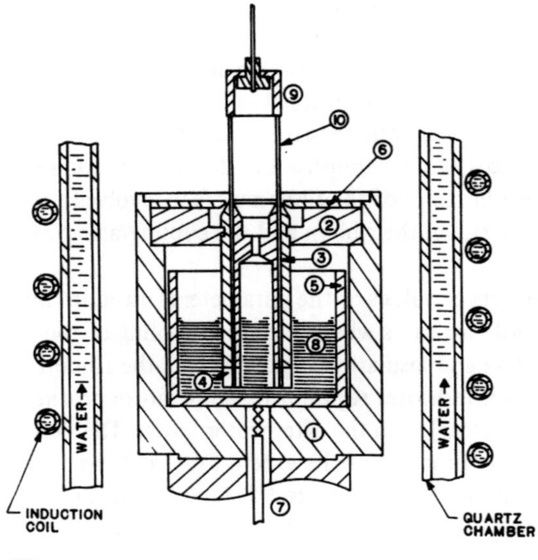

Fig. 17. Cross section of growth furnace and parts assembly used for the growth of small silicon tubes by EFG. *1* susceptor molybdenum, *2* crucible LID and DIE holder, *3* Die assembly (graphite), *4* feed capillary, *5* crucible (graphite), *6* thermal shield, *7* control thermocouple, *8* silicon melt, *9* seed, *10* new growth

The setup consists of a cup-shaped molybdenum susceptor which on the top holds a lid positioning the graphite die, which contains the feed capillary in the center of a silicon-filled graphite crucible. A simple thermal shield is placed near the die top, and the crucible temperature (the only external control element during growth) is in turn controlled by a thermocouple. (Of course, it should be clear from the foregoing discussion that the growth speed, which controls the release of the heat of fusion at the interface, can also be used as an element of control.) To complete the setup, the figure also shows the seed and a seed holder, as well as a piece of tubing already grown, both of which are connected to a vertical single-stroke pulling mechanism.

Figure 18 shows a photograph of such a growth station during operation, and the general simplicity of the design, which contains all relevant parts for EFG growth, is obvious. For the growth of single, up to 5 cm wide ribbons at rates around 2 cm/min, almost as simple setups suffice (Fig. 19), except that the thermal shielding (or insulation) is slightly more complicated, again due to the requirements of the edges. For long-term growth of ribbon, the belt-pulling device which is added to allow continuous withdrawal of ribbon from the top of the machine is clearly visible. Such simple machines have been operated reliably over the past two years by unskilled personnel on three shifts for the production of 2.5 cm wide ribbon at speeds around 2.0 cm/min. Growth is controlled here by the observation of the position of the two ribbon edges alternately through a binocular microscope. If the ribbon edges recede from the die edge, adjustments in the crucible temperature are made by the operator who also has to restart growth if, due to temperature fluctuations, freezing of the ribbon to the die occurs. In general, one operator is assigned to each growth station.

Fig. 18. Overall photograph of a tube growth station, showing induction heating supply on left; control console on right. In the center, the tube is clearly visible as it emerges from the heated region

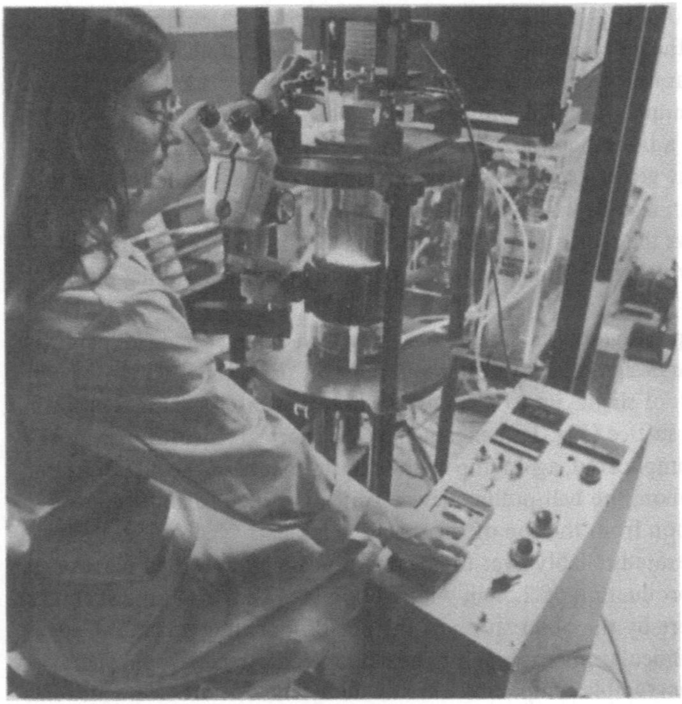

Fig. 19. Detail of modern induction heated growth station for the growth of 5 cm wide ribbon at speeds of ~2.0 cm/min. The power supply is not shown. On the right is the control console. In the center, above the induction coil, the passive afterheater arrangement is clearly visible. Also the ribbon is clearly visible in the center as it extends from the furnace into the lower end of the continuous belt pulling device

3.1 Multiple Ribbon Growth

The development of a multiple ribbon growth system was started in 1975 in response to initial economic analysis results[1)] which indicated that the output of ribbon per unit time and operator had to be drastically increased to significantly lower the cost of solar cell blanks. At that time, Tyco Laboratories produced routinely sapphire shapes grown in multiples from a single pull head. However, theory and experience by then indicated that the growth of silicon ribbon required far better control of the various growth parameters. Thus, the concept was evolved to base multiple silicon ribbon technology on linear arrays of individual ribbon growth units which were to share a common resistance-heated furnace to conserve power and floor space, reduce equipment cost, and provide for close spacing to ease operator access. Figure 20 shows a cross-sectional schematic of such a furnace, and Fig. 21 shows a photograph of the prototype that existed until the middle of 1979 with the capability of growing five ribbons, each 5 cm in width, and which has now been reconfigured to produce three ribbons, each 10 cm in width.

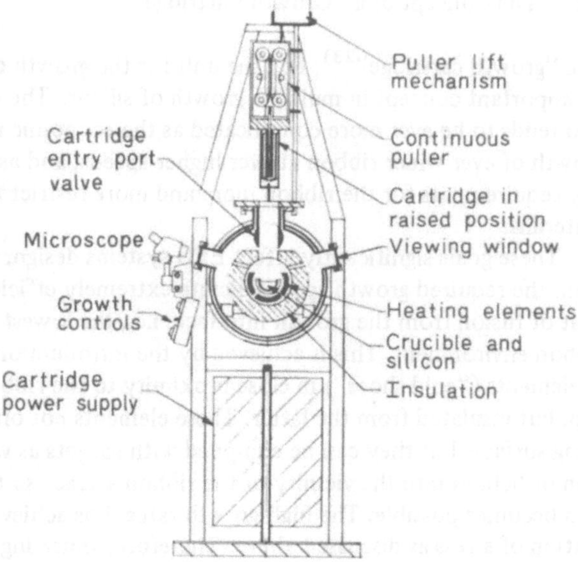

Fig. 20. Cross section of multiple ribbon growth furnace. (The major parts are indicated in the figure.)

Fig. 21. Full view of 5 ribbon multiple furnace. The five "growth cartridges" are completely lowered into the furnace. The "melt replenishment" unit, not yet completely lowered into the furnace, is visible on the left of the cartridges. The two panel racks house, in the main, a variety of diagnostic instruments

3.1.1 The Concept of a "Growth Cartridge"

The "growth cartridge"[23]), i.e., the unit for the growth of one ribbon, thus becomes an important concept in multiple growth of silicon. The execution of this concept also tends to be ever more complicated as the economic requirements dictate the growth of ever wider ribbon at ever higher speeds, and as the increasing chemical purity requirements for the ribbon more and more restrict the choice of construction materials.

These goals significantly affect EFG systems design. On the one hand, as we have seen, the required growth speeds dictate extremely efficient removal of the latent heat of fusion from the growth interface, i.e., the lowest feasible temperatures for the ribbon environment. This is achieved by the introduction of water-cooled heat removal elements ("cold shoes") in close proximity to the ribbon surface as well as the die top, but insulated from the latter. These elements not only provide a very cold radiating surface, but they can be equipped with gas jets as well, which allow introduction of helium into the vicinity of the ribbon surface so that convective heat removal also becomes possible. The high growth rates thus achieved, however, lead to the generation of stress as discussed above. Therefore, annealing afterheaters have to be provided. Also, die top temperature control through the use of the aforementioned face and end heater system is, of course, desirable particularly for the growth of wider ribbons, and these devices along with their power connections become part of the cartridge design. The requirements for high chemical purity in the ribbon, however, severely restrict the use of metal members anywhere in the cartridge where they cannot be shielded or held at very low temperatures. From all these various and often conflicting requirements evolved the successful present-day cartridge design.

3.1.2 Description of Cartridge

Figure 22 shows a schematic cross section of the growth cartridge with all the elements labeled[20, 23, 24]).

The cooling and heating elements for vertical and die top temperature control as well as the die are housed in an enclosure which is formed by four graphite walls and a graphite "floor". The walls are attached at the top of the cartridge to water-cooled stainless steel positioning blocks, which remain outside the furnace. The walls and the floor perform the important function of providing for critical alignment of the die top relative to all other components interior to the cartridge. These include (from the cartridge floor up) the die, graphite die heaters, die top radiation shields of either graphite or molybdenum, the (water-cooled) heat removal element, fabricated from molybdenum, nickel or stainless steel, a second set of radiation shields, and finally the linear cooling plates which contain an afterheater.

The growth slot, through which the ribbon is pulled, is centrally located between the linear cooling plates and the two halves of the heat removal element and is typically 0.15 cm wide. The distance between the die top and the maximum temperature point (lower edge) of the afterheater is approximately 2 cm. The steep temperature gradient in the solid near the growth interface which is necessary for high speed

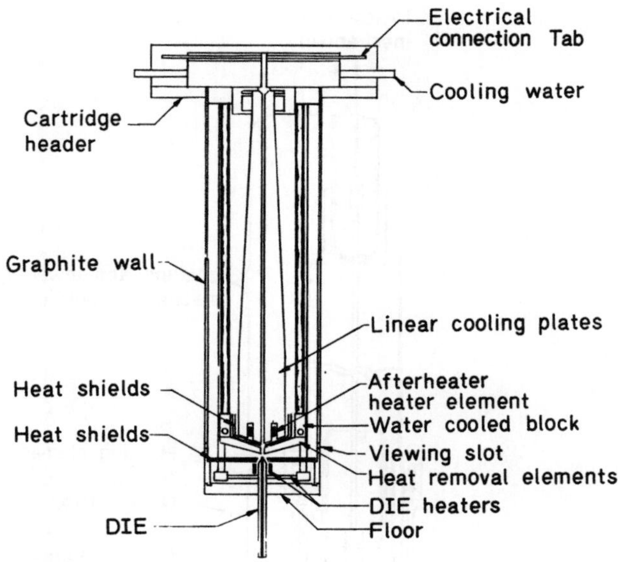

Fig. 22. Ribbon growth "Cartridge" – schematic. (The major parts are indicated in the figure.)

growth (~7.4 cm/min) is established by the heat removal elements, which, however, are somewhat counteracted by thermal leakage from the die heating elements which is minimized by the best possible shielding, provided by tightly fitting die-top radiation shields.

Ribbon leaving the space between the heat removal elements, as noted, has generally undergone the severe plastic deformation discussed previously (Sect. 2.3.2.1) because of the non-linear decrease in temperature above the growth interface. To relax the built-in stress as described (Sect. 2.3.2.2), the ribbon is reheated to the desired ~1100 °C to 1200 °C where fairly complete stress relief is possible during the time scale of the growth process, and it is then cooled in the required linear temperature profile, as it passes between the "linear cooling plates".

In the "normal" growth configuration, about 4 cm to 5 cm of the die extends above the melt level and into the cartridge, which leads to fairly large values of H_{eff} in these machines. Also in a cartridge, it would be obviously difficult to use just the crucible temperature to keep the die top near the melting point of silicon, and the melt would very likely be overheated in the process. This is avoided by the use of the separate die heaters, discussed in Sect. 2.3.2 (Fig. 13), and these also provide the major control elements for the shaping of the die top isotherms.

3.1.3 Melt Replenishment

The duty cycle of practically all crystal growth systems in use today is limited in part by the fact that continuous recharging of crucibles is either not practiced or fundamentally impossible for various reasons. In the present multiple EFG growth system,

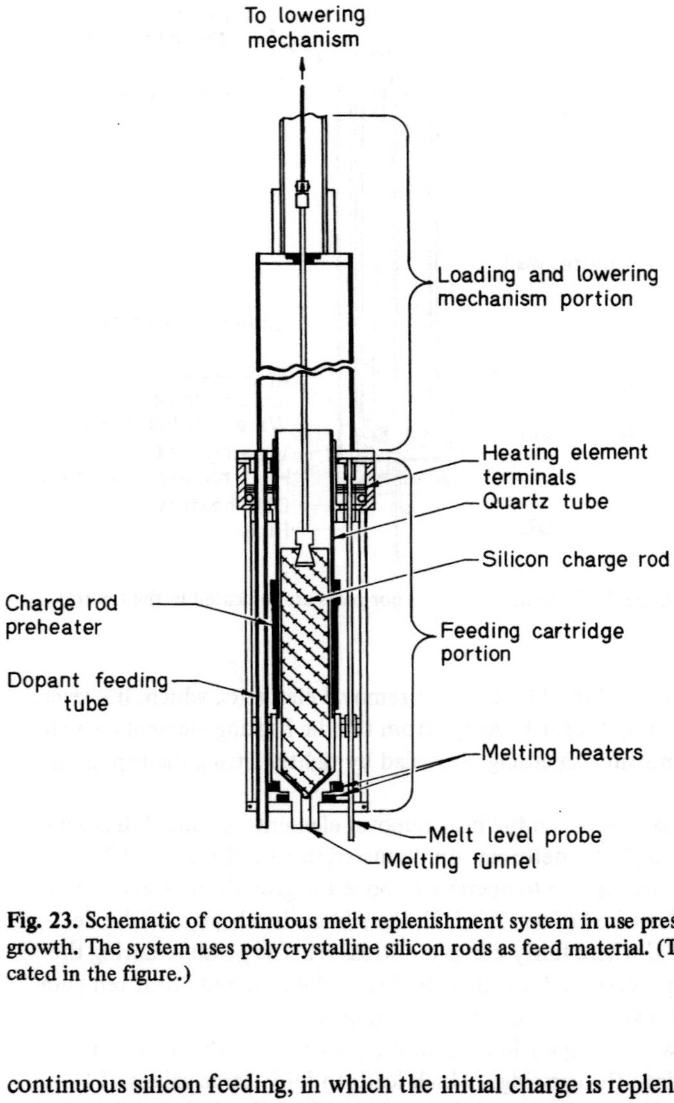

Fig. 23. Schematic of continuous melt replenishment system in use presently for multiple ribbon growth. The system uses polycrystalline silicon rods as feed material. (The major parts are indicated in the figure.)

continuous silicon feeding, in which the initial charge is replenished by silicon melted from a rod and allowed to fall into the melt as needed, is an integral feature of the overall design.

Figure 23[24] is a simplified drawing of the system used. All essential features of the silicon melting portion which is inserted into the furnace are shown, but the details of the lowering mechanism are omitted. These consist of a servomotor, gear train, and a length of subminiature roller chain from which the molybdenum support rod for the silicon charge is suspended. The system at present is designed to use as feed material polysilicon rods of 3.2 cm diameter by 25 cm length. At full operation of the furnace, approximately one charge rod is consumed in 30 min. The silicon rod is melted by contacting a heated "melting funnel" (made either of impermeable graphite or quartz). Two heating elements located near the point of melting and a preheater element surrounding the charge rod above the melting funnel supply

~1500 watts of total power to melt the rod. A graphite electrical probe senses the melt level. The signal from this probe is fed back to the lowering drive motor. The operator interface controls of this melt feeding system are designed to demand as little attention as possible; the only manual operation required is the occasional addition of new feed rods. A single up/down switch controls this function during normal operation.

Although this form of liquid melt replenishment has been found to be quite practical and reliable, it is of course not the only possible method. In fact, if silicon were available in small chips, pellets or the like, it would be simpler to recharge with solid material from a mechanism in which mechanical motion transports small amounts of solid silicon continuously into the liquid melt.

At the present time, however, highly pure silicon can easily be obtained in the form of rods, and the present melt replenishment system is thus a very practical solution.

4 Ribbon Quality Considerations

For the manufacture of solar cells at low cost, it is obviously essential that any material produced is sufficiently uniform in its shape and dimensions so that it can be processed by automated equipment. Since the kind of equipment to be used in future solar cell fabrication is not yet known, nothing specific can be said, but it may be safely assumed that specifications on the ribbon width, thickness and flatness will be fairly stringent, since these are important parameters in their effect on the yields in solar cell production.

However, any general discussion of "ribbon quality" is much more likely at present to refer to the question as to whether solar cells possessing high efficiencies can be manufactured from EFG grown ribbon. Unfortunately, and in spite of significant effort, no clear knowledge exists as yet regarding the relations between ribbon growth parameters, basic chemical and electronic properties, and finally solar cell efficiency. Hence, this is the most active field for investigation today, and much of what is mentioned below must therefore be phenomenological and fragmentary. Nevertheless, by empirical improvements, very satisfactory progress has been made. In fact, the top efficiency of a single, small EFG based cell produced at Mobil Tyco using very standard diffusion and metallization processes has been over 14%, and large lots of cells of ~25 cm^2 area (2.5 x 10 cm) have reached 11% average efficiencies in pilot production. However, yet higher efficiencies are still desirable, repreducibility of the product is also necessary, and the factors which prevent the achievement of such high efficiencies in wide ribbon ($\gtrsim$ 5 cm) grown at high speeds ($\gtrsim$ 3.5 cm/min) are also not understood at present.

All these questions are the subject of the discussion below which will accordingly deal with the factors affecting electronic quality of the material and also include a review of the approaches to geometric shape control and related problems such as flatness.

4.1 Electronic and Crystal Properties

There are a number of basic parameters which militate against the achievement of highly pure and crystallographically perfect material during EFG growth and thus against the realization of near intrinsic minority carrier properties. These parameters include:
(i) In EFG growth, the die isolates the bulk of the melt from the interface by imposing long capillaries through which back-diffusion is impossible. Hence, growth is thought to take place with an effective segregation coefficient of unity. Thus, all impurities drawn up the die must be incorporated in the solid ribbon.
(ii) The die, in intimate contact with the melt, has an important bearing on the quality of the grown ribbon, due both to the material from which it is fabricated and the geometry which it presents and which, consequently, controls the liquid-solid interface shape to a significant extent. Graphite, as noted before, has been used almost exclusively in silicon growth by EFG to date because of its good wetting behavior, its relatively low reactivity with silicon, its ease of fabrication into complex shapes, and its availability in highly purified form. It is, however, reactive with silicon to some extent and is, therefore, present in ribbon as a substitutional impurity and as silicon-carbide particles which precipitate and aggregate in the vicinity of the growth interface around the top of the die. These particles can be detrimental to growth stability by changing the local thermal distribution at the die top, and, when incorporated in the ribbon, they also often affect the cell performance.
(iii) In EFG, growth takes place at speeds about two orders of magnitude greater than in conventional melt growth of silicon. The effects of growth speed and cooling profile under these circumstances are not well known. In particular, the cooling/heating/cooling cycle induced by the cartridge systems at high growth speeds adds to the possibility that dislocations will be generated in very large numbers and may interact with impurities readily.
(iv) Finally, the complexities of a multiple cartridge system pose a threat to materials quality due to the large number and the variety of construction materials used. In many cases, it is not a trivial design problem to avoid metals which are well known to be detrimental impurities in silicon so far as solar cell performance is concerned. Thus, one is compelled to try and understand how these materials may be transported into the silicon during furnace operation and then, by design means, to prevent such transport. Also, many construction materials are not commonly available in highly purified form, and they may thus represent sources for unexpected impurities.

The following discussion will aim to present the current knowledge and opinions on the origin of the various defects, their electrical manifestation and possible means to their prevention; unfortunately, the fragmentary nature of all present knowledge in this area will thereby become quite apparent.

4.1.1 Crystal Defects in EFG Ribbon

EFG ribbons commonly contain a variety of crystallographic defects including dislocation densities up to $10^6/cm^2$ (but mostly around $\sim 10^4/cm^2$), dislocation arrays, intersecting twins, stacking faults, and, as most prominent features, silicon-carbide particles and planar defects which have been identified as pairs or bundles of twins. All of these defects have been shown to be electrically active to some extent[25, 26], i.e., they locally reduce the minority carrier lifetime. In most cases, however, there are strong reasons to believe that these defects in themselves would not be nearly as detrimental as they appear, if they did not act as sinks for chemical impurities which are present in the ribbon[27]. Thus, efforts to improve the material with respect to solar cell efficiency today are equally aimed at the reduction in defect levels, and improvements in the chemical quality of the ribbon, as well as with possible beneficial interactions among impurities and defects, i.e., internal gettering, complexing, and the like.

Also, in the course of recent improvements in the growth process, crystallographic quality of the ribbon has been steadily increasing. The occurrence of grain boundaries, both high and low angle, intersecting twins, and stacking faults has become more rare as the growth process is being better controlled, and it seems proper at this time to consider them aberrations, rather than inherent features of EFG ribbon. Although it is quite clear that their occurrence must be minimized, they are not further discussed here.

The discussion following will focus attention on the three types of "generic" defects which at this point appear to be an inherent and inevitable consequence of the preparation of silicon by EFG with carbon dies.

4.1.1.1 The "Steady-State" Defect Structure

EFG grown ribbons, when etched, show as their most noticeable morphological feature arrays of "linear" defects. In the ideal, these defects are all parallel to each other (Fig. 24), parallel to the growth direction, and extend through the full thickness of the ribbon (Fig. 25). The ribbon orientation in these cases has been determined as $\{110\}\langle 211\rangle$[28].

On closer examination, the lines seen are determined to be twin pairs or bundles of twins[26–29]. According to recent work by Ast and Strunk[30], the twin pairs are of a fairly regular nature, though rather closely spaced, but the "irregular" twin bundles of exceedingly close spacing which are observed are unique to EFG of ribbons (Fig. 26).

It was shown[28] that the structure is almost immediately generated, if single crystalline seeds of $\{110\}\langle 211\rangle$ orientation are used, or if ribbons already containing the defect structure are employed as seeds. However, when for instance $\{111\}\langle 110\rangle$ seeds are chosen, it may take about 30 cm of growth to obtain full alignment of the twinned structure with the growth direction.

Although this structure is quite prominent, the mechanism of its formation is still in dispute[14, 31, 32], but there is nearly universal agreement that it is very probably related to the presence of carbon at the solid-liquid interface and in the grown ribbon, which in turn is a direct consequence of the use of a carbon die.

Most recently, a speculative model that has been presented[32] postulates the presence of random local perturbations in meniscus height and the attendant formation

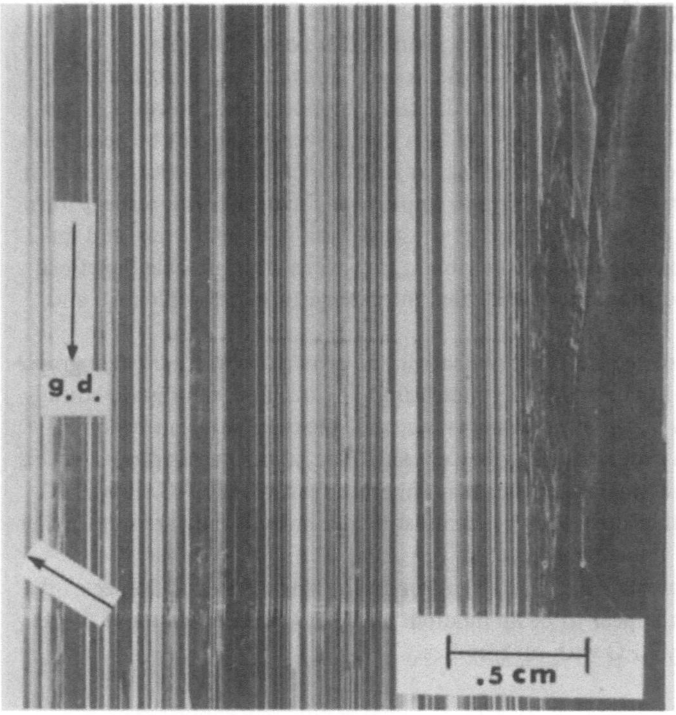

Fig. 24. Optical micrograph of the surface of an etched silicon ribbon showing the "linear boundaries" of the equilibrium defect structure. (g.d. indicates: growth direction)

of cusps which contain high carbon and impurity concentrations and thus are stabilized by the increased infrared absorption which would ensue. The paper also shows some evidence for cusp formation, which can be further substantiated by experiments

Fig. 25. Cross sectional micrograph (through the ribbon thickness) showing "linear boundaries" of the equilibrium defect structure extending through the full (~ 250 μm) ribbon thickness

Fig. 26. High resolution lattice image of microtwins in EFG ribbons. Viewing direction is along [110]. The upper half shows two coherent microtwins which are 4 and 11 (111) planes thick. The distance between the major twin lines is thus ~ 50 Å. The lower half shows an incoherent microtwin on a (112) plane running from the top (left side) to the bottom (right side) of the thinned sample. (Courtesy of D. Ast and H. J. Strunk, Cornell University)

in which the solid ribbon is quickly withdrawn from the liquid (Fig. 27)[33]. Recently, it was also suggested[34] that cusp formation could follow quite naturally, if growth were to occur in a two-phase mixture of Si and Si–SiC eutectic, i.e., if the liquid was not "carbon saturated". Here, the Si–SiC eutectic (although its melting point has been reported lie only ~0.2 °C below that of Si)[35] would still be liquid significantly ahead of the main growth front in the high gradient of EFG growth because of its lower

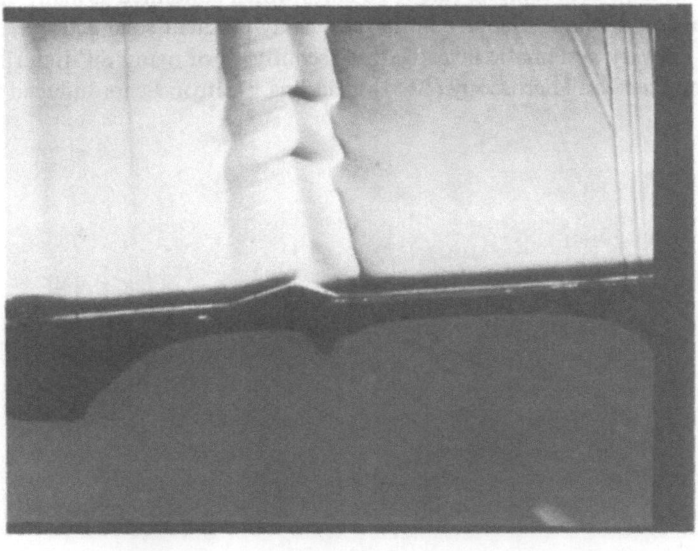

Fig. 27. "Cusp" formation at the growth front of an EFG ribbon which is quickly withdrawn from the melt. A "twin" boundary apparently forms at the center of the "cusp". Magnification 70 ×

freezing temperature. Undersaturation in carbon content could, of course, be a consequence of the high growth speeds and an attendant failure of the silicon to come to equilibrium with the carbon die surface when grown from quartz crucibles. Also, recent data on the Si–C–O equilibrium[36)] suggest that decarburization can occur (even when a carbon crucible is used to contain the silicon melt), if, as in most practical cases, oxygen is present in the growth system.

However, none of the models thus far have been definitely proven to be correct, and, furthermore, it has been shown that in general the "linear" defects have only a moderate influence on the recombination in silicon[25)]. Certain regions of them can, however, be shown to reduce current sollection more than others[37, 38)] which may indicate that perhaps the more complex twin bundles could act as impurity sinks. Since solar cell efficiencies higher than those obtained commonly today are sought, the question of eliminating the "linear" boundaries is still a valid one.

4.1.1.2 SiC Particles: Origin, Effects and Elimination

All early EFG grown ribbons, as well as most any which are grown at present from new die designs, new growth setups, or under generally non-optimized conditions, are observed to be studded with small particles (Fig. 28), just visible to the naked eye, which can easily be shown to consist of SiC. They are indeed a perturbing presence as it is readily apparent that many of them constitute serious shunting paths for solar cells[25)]. Moreover, several investigations also revealed anomalies related to their behavior. Many of these particles seriously perturb the lattice around them, leading to halos of what presumably are impurityrich zones which have severely degraded minority carrier collection properties, while others are found not to be detrimental. In some cases, large area samples (2.5 x 10 cm^2) which contained as many as ten such particles could be manufactured into reasonably efficient solar cells[39)].

A very systematic investigation recently[40)] of many SiC-rich ribbons using Scanning Electron Microscopy (SEM) including Electron Beam Induced Current (EBIC)

Fig. 28. 5 x 10 cm EFG silicon ribbon section with visible surface inclusions of SiC

measurements, shows that it is possible to sort particles into four general classes which can be distinguished by the recombination behavior of their surroundings. According to these observations: Category 1 SiC particles are fairly large crystals up to 300 μm in size which produce little or no detectable recombination, as well as little or no structural disturbance in the ribbon surrounding them. These particles, which constitute about 25% of all those studied, also are apparently not deeply embedded in the ribbon; they rather look superficial.

Into Category 2, the authors put about 10% of all particles which they find to be small, ~30 μm, and to occur singly or in groups. They produce little or no visible grain structure (but may be surrounded by dislocation clusters) in the ribbon. They are deeply embedded and lead to considerable recombination up to several hundred μm's away. The recombination pattern is described to be radially symmetrical in this case.

Category 3 consists of groups of intergrowths of comparatively large particles (~150 μm in size) which cause considerable structural disturbances in the ribbon, as well as significant recombination. Here, the recombination pattern is elongated in the direction of growth. These particles, again, are deeply embedded in the ribbon and occur with the highest frequency; about 45% of all particles in the ribbons studied belong into this group.

Finally, in Category 4 are moderately sized SiC crystals 40 μm to 50 μm large which are not as deeply embedded as those in Category 2 or 3, but more so than Category 1. Little or no structural disturbance is produced, but a considerable increase in recombination without any spatial feature occurs around the particle.

These investigations clearly show the complexity of the phenomena caused by SiC particles when they occur, and they highlight the necessity to prevent their occurrence.

However, there is today no clear understanding as to how these particles are generated during growth, except that one may suggest that the least offensive Category 1 is perhaps due to gas phase reactions. The others have to be presumed to be generated during the growth process, and a number of observations can be made to this effect. One can, for instance, observe that during growth SiC particles which are visible on the die top are regularly picked up by the moving ribbon and incorporated into its surface.

Moreover, when the meniscus height momentarily goes to zero, i.e., the ribbon "freezes" to the die top, rows of SiC particles cling to the ribbon. Indeed, a skilled operator can, through a downward adjustment of the growth speed, make this happen with regularity and so clean the die, causing the subsequent growth of SiC-free ribbon for a short time. Further, when growth is very stable, it is possible to observe that particles are preferably picked up at the ends of the ribbon where the meniscus is low. On occasion, however, one also observes the occurrence of a particle on the moving ribbon, without seeing one obviously dislodged from the die top, even if two observers view both sides of ribbon and die simultaneously.

In general, these kinds of observation, along with the assumption that carbon must be rejected at the solid-liquid interface because of its reported k_{eff} of 0.07[41)], have given rise to the general hypothesis[42)] that, in the diffusion boundary layer, SiC will be nucleated. These nuclei then would either fall to the die top and grow, if their mass allows them to do so, or they would grow near the interface until they can fall to the die top through the moving liquid in the meniscus. Here, they will further grow to a size which eventually is sufficient to bridge the gap between the die top and the solid silicon at the interface. They will then be gripped by the moving ribbon and be torn off the die top. Even though this hypothesis was never explicitly proven, it can explain a sufficient number of observations to become a logical starting point for experimental designs.

One would thus predict that the creation of a very high meniscus should lead to the elimination of SiC, at least for a certain length of growth. Indeed, the success of Ciszek (this volume), who produced relatively SiC-free ribbon using high menisci, might be taken as proof for this hypothesis. However, one would of course also predict that during long periods of growth the SiC particles would eventually still become large enough to bridge the meniscus. However, it is generally not possible to test this by observation over long periods of growth, which has indeed been carried out for several days, because it is exceedingly difficult to keep the meniscus height constant at all times and to avoid "freezes". Thus, a random distribution of particle sizes still results from such experiments, and the question posed by the original hypothesis and the experiments of Ciszek is therefore as yet unanswered.

In the meantime, another remedy to the SiC problem has been found to be effective in a number of situations[43)]. It was observed that elevating one side of a die over the other by as little as ~0.05 to 0.1 mm (Fig. 29) causes the complete disappearance of SiC on one side of the ribbon, even in growth runs in which ~30 m of ribbon is grown over 15 hours. The effect is not quite clear; one may though perhaps ascribe it as follows. Since one side of the die is higher than the other, the meniscus is lower (Fig. 30). Hence, SiC particles will preferably precipitate there with a high frequency. In consequence, particles on the high meniscus side are retarded in growth due to lack of carbon. In effect, the high frequency of precipitation on the low meniscus side causes the carbon concentration on the high meniscus side to be below saturation.

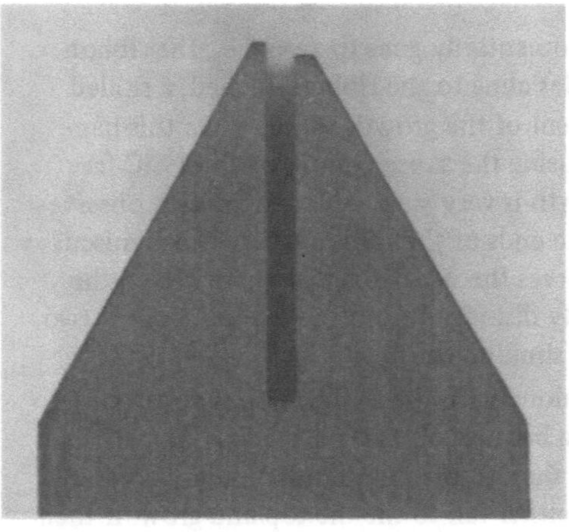

Fig. 29. Photograph of the cross-sectioned upper half of a "two-piece", "displaced", graphite EFG die. The right hand edge is clearly lower, i.e., "displaced" with respect to the left. The "top slot" capillary dimension is 500 μm

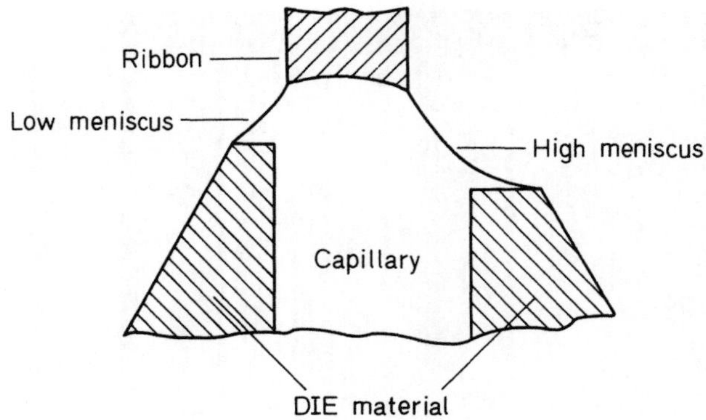

Fig. 30. Schematic of the assumed meniscus shape when using "displaced" dies. SiC is only deposited on that side of the ribbon which grows from the higher die side

This hypothesis suffers somewhat from the observation that on many occasions the SiC density, even on the low meniscus side, is not especially high. Hence, other factors which control the SiC density are apparently also operative. So far, the "displaced die" is the only remedy which avoids SiC particle generation on one side of the ribbon totally, even in long growth runs, at least when moderately high menisci are used and reasonably steady-state growth is achieved.

Nevertheless, in any situation where new die designs or growth conditions are necessary, as in the growth of wider ribbons or growth at higher speeds or from machines with different thermal conditions, SiC particle density will be a parameter which requires particular attention. Unfortunately still, empirical experimentation at this point appears indispensable.

4.1.1.3 Dislocations

Although the major defect types on which attention has been focused are the above-mentioned micro-twins and the silicon-carbide particles, high dislocation densities are also a recurring general feature of EFG ribbons[14, 26]. Dislocations often appear singly or in arrays when the samples are etched (Fig. 31a,b); they are also found bunched around SiC particles or piled up against grain boundaries.

The origin of these various dislocation morphologies is not known; in particular, it is not clear whether they are generated during the growth process or during the rapid cooling of the ribbon, particularly during cartridge growth where the heating/cooling/heating cycle described previously is employed.

However, it has been shown that the dislocation bunches found around SiC particles and the linear arrays of dislocations often found are particularly harmful as they cause significant reductions in the minority carrier lifetimes.

In general, the dislocations which have so far been discussed have all been found by observing etch pit densities. Recent investigations by Ast[44] using transmission electron microscopy have revealed, though, that the dislocation density may be high-

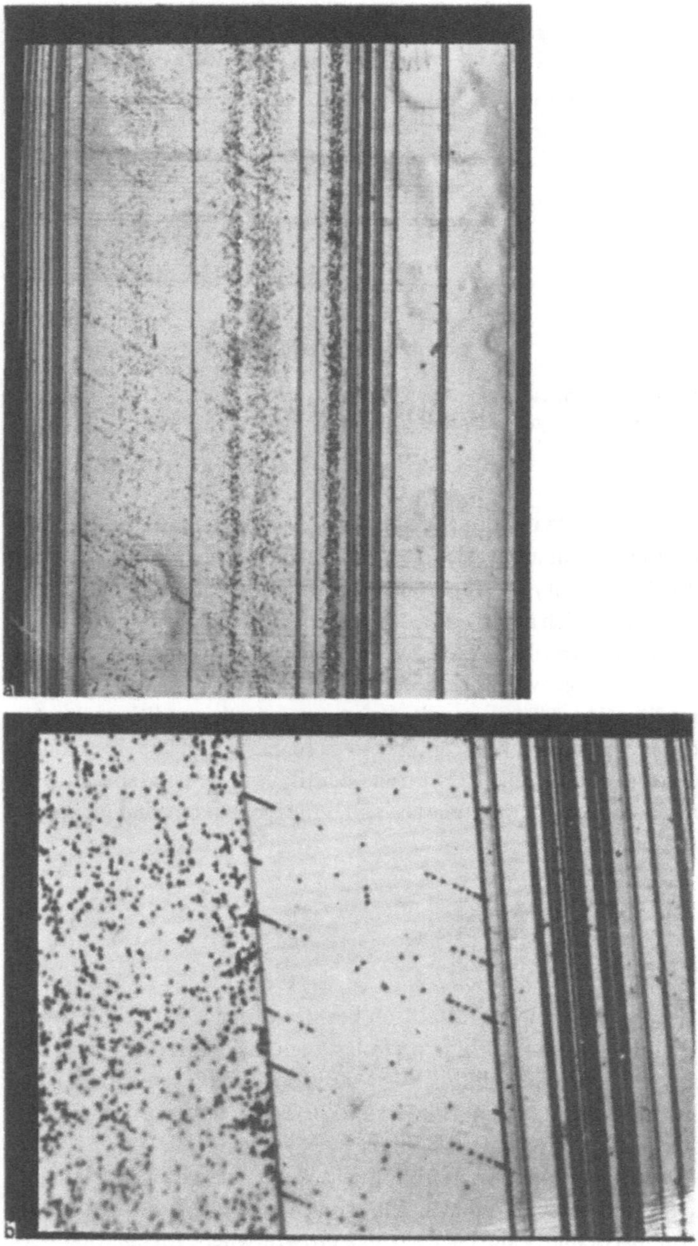

Fig. 31 a, b. Examples of "linear boundaries" and dislocations, found in EFG ribbon
a Magnification 72.5 X. Various "bunched" dislocations are visible,
b Magnification 262.5 X. Arrayed dislocations are clearly visible

er than the etch pit densities suggest, as it was found that the usual etches did not always uncover all the dislocations present.

No particular growth configuration which reduces dislocations during EFG growth has as yet been suggested, but it is generally observed that, with the evolutionary improvements in crystalline quality which have been achieved in EFG ribbons over the past five years, the general dislocation density is now at reasonable levels (10^3 to $10^4/cm^2$). Here again however, further reductions may have to be sought, if large increases in solar cell efficiencies over those obtained now are to be realized.

4.1.2 The Role of Chemical Impurities

The concern about chemical impurities, as pointed out above, rests of course in the fact that one does not expect any purification to occur during EFG. Thus, all impurities which are contributed by the starting material, the crucible, the die, and the general apparatus, are likely to be found in the solid-shaped crystal. This expectation has recently been confirmed experimentally for the case of Al[45)].

However, various growth conditions may lead to distinctly different distributions of the impurities through the volume of the ribbon. It has for instance been shown [45, 46)] that particular die designs can cause inhomogeneous distributions of impurities over the ribbon width. Also, the recombination patterns around the various classes of SiC particles mentioned above (Sec. 4.1.1.2) suggest rather different impurity distributions[40)]. On occasion, too, grain structures develop through the center portions of thick ribbons grown at high rates, and impurity accumulation is suspected there[32)]. On a more microscopic scale, it is observed that crystallographically identical defects may not cause severe reductions in local lifetime[37, 38)] as they are observed in Electron Beam Induced Current measurements (EBIC) or generate inhomogeneities in local resistivity as determined by spreading resistance studies.

Thus, one important aspect of impurities in EFG ribbon is that they apparently accumulate in high concentrations locally[40)]. This can, of course, lead to severe local degradation in the recombination properties of the material. On the other hand, one can invoke an "internal gettering" argument and suggest the possibility that under the right conditions impurities could be rendered harmless by complexing or clustering with defects, or outright precipitation in the form of insoluble and highly resistive compounds. What militates against the latter explanation is the general absence of any precipitates in the ribbon, even when it is studied in high resolution Transmission Electron Microscopy (TEM)[30, 47)]. Thus, aside from the obvious impurity accumulation around SiC particles as mentioned above, attention has always been centered on the twins and twin bundles of the "steady-state" structure as the major impurity sink. So far, however, it has been impossible to show that impurities reside there in electrically active or inactive states, or even to definitely demonstrate that impurities and not the defects themselves are responsible for the electrical activity seen in some regions by EBIC.

Thus, at the present time it is somewhat puzzling why EFG ribbons which are always grown under assumed $k_{eff} = 1$ conditions can possess quite acceptable diffusion lengths, in the range of 30 to 70 μm. So far, employing relatively sophisticated chemical analysis methods including ion microprobing, SIMS, spark source mass spectroscopy and direct nuclear reaction techniques as well as others has not provided much insight into this vexing question.

4.2 Geometrical Considerations

Width, thickness, and flatness of ribbons are quite important parameters which affect the yield in solar cell manufacturing, and which therefore must be closely controlled.

4.2.1 Thickness and Width Control

From the discussions in Sect. 2, it became quite apparent that the thickness control of a ribbon during EFG is a direct function of the die top dimensions, H_{eff} and the meniscus height, the latter of which in turn is controlled by the die top temperature, as it is discussed in detail for tube growth in Sect. 2.3.1. Thus, if one observes the meniscus height, it should be possible to keep it constant with a suitable control loop which adjusts either the die top temperature or the pulling speed.

The dynamics of such a control process and very practical means to its implementation have been recently investigated in detail[48]. However, a similarly detailed and fundamental understanding of the dynamic processes operative at the ribbon edges is not yet available (see Sect. 2.3.2). It appears quite possible, though, to use similar means of observation, as they are detailed in Ref. 48), along with adjustments of end heater power or speed control devices as mentioned, to hold the ribbon width within fairly close tolerances, particularly when the process is aided by die designs which minimize edge movement.

4.2.2 Ribbon Buckling

In a previous section (2.3.2.2) a solution, in the form of a proper afterheater arrangement, is provided which allows growth of stressfree, flat ribbon at speeds up to ~5 cm/min. Above this speed, another phenomenon is observed[20, 24, 49] which appears to require further analysis of thermal stresses and very likely additional refinements in cooling profiles. This phenomenon is the generation of stress-induced buckles of rather large amplitude which make the ribbon completely useless for solar cell fabrication. Buckling apparently occurs during growth to relieve compressive stresses acting over certain lengths of ribbon. These are stresses left by incomplete relief in the region below 1200 °C in a nonuniform gradient. The buckling tendency increases for wider and thinner ribbons and when longer ribbon segments are in compression. Fundamentally then, buckling is believed to be caused by the incomplete stress relief in the heating/cooling/heating cycle which is a result of the present cartridge temperature profiles[20].

In addition, the alignment of the mechanical parts which guide the ribbon from the die top appears perhaps to play a major role in causing non-flatness of ribbon, and the possible interactions between the guidance system and the "stress buckling", which may increase the degree of non-flatness, must therefore also be considered.

These questions are not at present under much active investigation, because growth speed increases much over 4 cm/min are not now considered a high priority. Nevertheless, these problems still exist, they have neither been completely explained, nor are there empirical solutions to them available. It is quite clear, however, that these phenomena will become of paramount importance if growth speeds much over 5 cm/min are sought[20], particularly for wide (> 5 cm) ribbons.

4.3 Solar Cell Efficiencies

The general discussion of growth defects in EFG ribbon and the complications which may arise from the $k_{eff} = 1$ condition, which is obtained during EFG, has always given rise to the conclusion that the lifetimes and thus the solar cell efficiencies of the resulting material would be limited to relatively low values. However, through the recent years of ribbon development these values have steadily increased to a level which makes the material quite practical for terrestrial solar cell applications. In fact, the highest values (Table 3) which have been achieved on small area cells prepared by standard processing methods[50] from parts of ribbons grown at ~2 cm/min in RF heated systems are not significantly lower than those of cells prepared from Czochralski grown silicon with similar solar cell preparation processes. Also, through evolutionary developments in both crystal growth of 2.5 cm wide ribbons (including the refinement of graphite parts purification procedures[29]) and device processing of such materials, significant advances in average efficiencies of large area cells have been achieved in the past several years.

Induction heated systems, in three-shift operation, produced single ribbons at rates around 2 cm/min which, when processed into cells of 2.5 x 10 cm^2 areas by standard methods, have often reached efficiencies of 11+% (AM1) in large batches (Table 4), with averages over a whole year reaching ~10%.

Nevertheless, the question as to why such cell efficiencies cannot be routinely achieved on materials grown from the more modern induction or resistance heated

Table 3. Performance of Selected Solar Cells from Ribbon Grown in Induction-Heated Growth Equipment. Grown at ~2.0 cm/min, 2.5 cm wide; Light Intensity, 94.5 mW/cm^2; ELH Lamp, 28 °C

Area (cm^2)	V_{oc} (mV)	I_{sc} (mA/cm^2)	FF	η (%)
4.20	564	29.8	0.730	13.0
4.26	567	29.1	0.737	12.9
4.06	580	29.6	0.774	14.1
4.25	574	29.9	0.742	13.5
4.13	562	30.0	0.736	13.6

Table 4. Solar Cell Efficiencies of a Large Batch of 2.5 x 10 cm^2 Cells Prepared from One Ribbon, 2.5 cm Wide, Grown at ~2 cm/min from Simple Induction-Heated EFG Equipment

Cell No.	Area (cm^2)	I_{rv} (mA/cm^2)	V_{oc} (V)	I_p (mA)	I_{sc} (mA/cm^2)	FF	P (mW/cm^2)
19948	21.74	0.090	0.558	556.3	27.37	0.737	11.23
19949	21.74	0.224	0.560	519.7	27.66	0.733	11.34
19950	21.74	0.449	0.554	519.1	26.80	0.728	10.88
19951	21.74	0.471	0.562	550.8	28.37	0.720	11.47
19952	21.74	0.247	0.566	530.1	27.90	0.731	11.54
19953	21.74	0.090	0.568	548.3	28.99	0.718	11.83
19954	21.74	0.133	0.570	530.0	28.07	0.740	11.84
19955	21.74	0.202	0.561	534.0	27.45	0.730	11.26
19956	21.74	0.112	0.575	571.9	28.36	0.752	12.27
19957	21.74	0.247	0.564	548.8	28.52	0.693	11.14
19958	21.74	0.269	0.552	509.0	26.78	0.716	10.59
19959	21.74	0.157	0.569	573.1	28.45	0.747	12.09
19960	21.74	0.314	0.563	539.1	28.24	0.732	11.64
19961	21.74	0.247	0.566	558.1	29.43	0.736	11.85
19962	21.74	0.269	0.562	529.3	27.74	0.724	11.28
19963	21.74	0.135	0.568	552.1	28.06	0.739	11.77
19964	21.74	0.247	0.556	521.7	27.48	0.712	10.87
19965	21.74	0.202	0.564	573.5	27.76	0.726	11.37
19966	21.74	1.863	0.560	429.9	23.64	0.695	9.20
19967	21.74	0.247	0.569	537.1	28.17	0.751	12.04
19968	21.74	0.090	0.561	539.7	27.74	0.733	11.40
19969	21.74	14.770	0.529	300.4	26.06	0.386	5.31
19970	21.74	0.314	0.572	581.7	28.89	0.750	12.40
19971	21.74	0.202	0.560	527.7	27.55	0.717	11.06
19972	21.74	0.471	0.564	541.2	27.82	0.727	11.40
19973	21.74	0.404	0.560	531.8	27.75	0.731	11.35
19974	21.74	0.180	0.567	549.6	27.93	0.746	11.81
19975	21.74	0.022	0.568	572.6	28.24	0.753	12.08
19976	21.74	1.998	0.567	532.2	28.10	0.716	11.45
19977	21.74	0.606	0.566	535.5	28.74	0.665	10.82
19978	21.74	0.292	0.563	546.7	28.24	0.704	11.20
19979	21.74	1.437	0.572	572.3	29.19	0.711	11.86
19980	21.74	0.202	0.568	541.4	28.71	0.711	11.60
19981	21.74	0.516	0.560	551.6	20.11	0.712	11.21
19982	21.74	0.718	0.568	552.3	28.76	0.717	11.71
19983	21.74	0.098	0.561	540.4	28.07	0.724	11.41
19984	21.74	0.224	0.563	563.8	28.50	0.724	11.62
19985	21.74	0.494	0.553	521.8	27.27	0.722	10.08
19986	21.74	0.471	0.557	537.3	27.55	0.717	11.02
19987	21.67	0.473	0.563	527.5	27.62	0.713	11.09
19988	21.74	0.965	0.565	545.7	28.68	0.706	11.44
19989	21.74	0.112	0.569	569.0	27.06	0.728	12.04

Table 4. (continued)

Cell No.	Area (cm^2)	I_{rv} (mA/cm^2)	V_{oc} (V)	I_p (mA)	I_{sc} (mA/cm^2)	FF	P (mW/cm^2)
19990	21.74	0.673	0.555	535.5	27.75	0.721	11.12
19991	21.74	0.292	0.555	519.6	27.82	0.695	10.73
19992	21.74	0.898	0.563	546.1	28.25	0.704	11.20
19993	21.74	0.337	0.563	539.9	28.19	0.712	11.29
19994	18.68	2.038	0.561	419.2	25.95	0.690	10.04
19995	16.11	0.151	0.554	356.4	25.05	0.724	10.03
Mean Value	21.56	0.760	0.562	530.7	27.83	0.713	11.21
Std Error of Mean		0.3059	0.0010		0.145	0.0073	0.151

growth stations which produce wider ribbons at higher growth speeds cannot as yet be conclusively answered. Significant differences between materials which are produced from growth equipment of slightly varied design and under somewhat differing growth conditions are observed, and these materials often even differ in their response to differing cell processing conditions.

The reasons for these differences are only slowly becoming clearer with a great deal of active investigation at this time, and it is not possible, therefore, to deal with the details of the current state-of-the-art in this paper. Rather, the reader is directed to the original literature which presently deals with various aspects of this topic[51–56]. However, it is instructive to consult Fig. 32 which plots the progress of the efficiency in very large area cells over the recent two years of operation of a resistance heated cartridge furnace in which a systematic effort was undertaken to study the response of the electronic properties to certain equipment and growth variables[57].

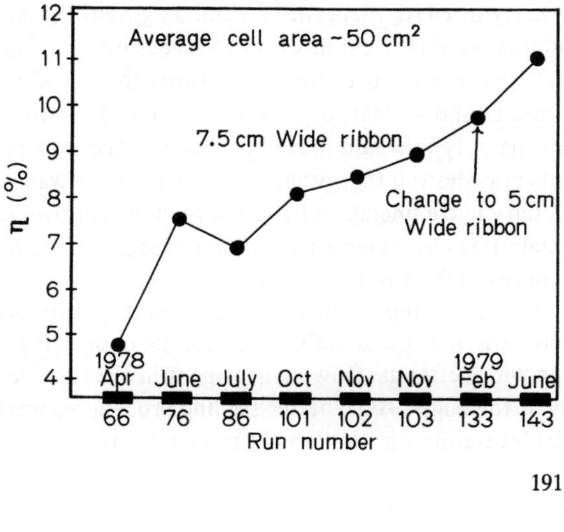

Fig. 32. Progress over time in the efficiency of large area solar cells prepared from "cartridge" type growth equipment

The early improvements in efficiency here are associated with a change in cell processing when it was discovered that these materials responded better to a CVD diffusion process over 1000 °C[58] than to the standard phosphine diffusion[50]. Also, a certain improvement can be traced to changes in die design in which multiple distributed capillaries were replaced with large, centrally located capillary slots[20].

The next noticeable change in the upward slope occurred in October 1978 with the introduction of the displaced die[43] which led to the preparation of ribbon which was completely free of SiC on the junction side[57]. The most recent work, which produced large area cells of 11+%, included changes of the furnace inert gas purge rates[57] which are not yet completely understood, particularly since these ribbons are also grown at lower speeds (~2 cm/min). However, it is clear that further progress in EFG growth and processing technology, as well as a further understanding of the fundamental connections between growth and processing variables and electronic properties of the ribbon, should lead to large area cell efficiencies which approach or exceed those shown in Table 3.

5 Present Status of Multiple Ribbon Growth

As noted in the introduction, it was an early assumption[1] that EFG of silicon ribbon would allow high throughputs and thus low cost, not only by achieving large ribbon widths at high growth rates, but also by allowing multiple ribbon growth from single pieces of equipment. In response to this, the growth station described in Sect. 3.2 was developed as a laboratory demonstration unit which would allow testing of the design concepts during long growth runs. Thus, stability of growth and subsequent performance of the grown material as a solar cell could be determined and the mechanical integrity and the overall duty cycle of the equipment could be assessed. Finally, such a laboratory test would serve to supply better input data for a refined economic analysis, based on the well-developed SAMICS[59] concept, which requires significant details of machine operation and construction in order to yield useful results[24].

Early in 1978, then, the five ribbon growth station, shown in Sect. 3.2, which produces material 5 cm in width at speeds up to 3.5 cm/min, had been developed to a sufficient degree to attempt initial growth runs which are described in Ref. 23) and it could be shown that, by making reasonable assumptions, an advanced system of this general type would indeed produce ribbon at very low cost. These early experiments also showed that progress in several areas was necessary to establish routine high duty cycle operation in such a growth unit over long periods of time as well as to guarantee cleanliness of the final product, before the design could be considered prototypical for a production unit.

A series of improvements was, therefore, introduced[57]. These include the introduction of a special bulbous-ended die design (Fig. 33) to aid ribbon edge stability along with better gas flow management in the furnace to enhance stability, particularly at the edges. Also, better seeding procedures were developed. A high degree of cartridge standardization was introduced, and the melt replenishment unit was re-

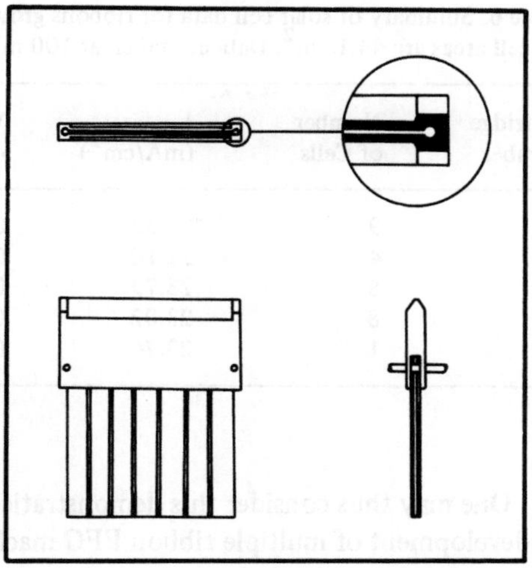

Fig. 33. Schematic section drawing of a "bulbous ended" two piece die. The "bulbous" end serves to stabilize the ribbon edge against "tapering" and thus contributes significantly to geometrical stability. Also, "pinning" of upper to lower die section is clearly visible

designed. Finally, an almost complete carbon-based design for the furnace interior was developed to minimize melt contamination, and "displaced dies"[43] were used to manage SiC particle density.

These improvements now permitted highly successful demonstration runs to be made which qualify the general design of the furnace as prototypical for future, larger production units. As shown in Table 5, a total of 150 m of ribbon was grown in a 15.5-h period at a duty rate of ~94%, and the conversion efficiency of solar cells fabricated from this ribbon was close to 9% (Table 6).

Table 5. Multiple-ribbon throughput data for 15.5-h growth demonstration run, May 21, 1979

Cartridge No.	1	2	3	4	5
Total Quantity (m)	30.4	29.6	29.9	31.1	27.7
Total Duration of Growth (min)	10	890	825	919	829
Percentage of 15.5-h Run Period Actually Growing	97.8	95.7	88.7	98.8	89.1
Number of Freezes	3	5	6	3	4
Longest Duration of Continuous Growth (min)	692	331	505	490	508
Average Growth Rate (cm/min)	3.34	3.33	3.62	3.38	3.34
Overall Duty Rate (%)			94.7		

Table 6. Summary of solar cell data for ribbons grown in multiple-ribbon growth run 16–187. All cell areas are 44.1 cm^2. Data are taken at 100 mW/cm^2; ELH light, 28 °C; cells are AR-coated

Cartridge Number	Number of Cells	J_{sc} (mA/cm^2)	V_{oc} (V)	FF	η (%)
1	9	23.32	0.534	0.696	8.68
2	4	22.10	0.520	0.698	8.02
3	5	23.79	0.537	0.714	9.12
4	8	23.02	0.530	0.700	8.55
5	1	23.74	0.534	0.672	8.52

One may thus consider this demonstration run to accurately represent the state of development of multiple ribbon EFG machines. In most respects, the system appears to be quite reliable, and operating procedures are well defined. The system could, in fact, be run on a routine basis requiring only a few detail refinements. It would then produce ~100,000 cm^2 of silicon in 24 hs, and at 9% device efficiency, that would amount to ~900 peak watts of solar cells per day. Since power consumption of the furnace is ~32 kW, the energy payback time for the sheet growth process itself would be ~850 hs.

However, throughput parameters derived from the previous SAMICS study[24)] of multiple EFG suggest that further advancements in throughput and efficiency are required to decrease the add-on costs of the sheet growth step below $20/m^2 which is estimated to be necessary to meet required photovoltaic systems prices which allow reasonable electric power costs[60)].

The multiple growth station is, therefore, being developed further to allow multiple growth of ribbons 10 cm in width at speeds $\geqq$ 4 cm/min.

6 Conclusions and Future Outlook

It is quite apparent that presently the effects of capillarity as they relate to EFG of silicon are quite well understood. The principles derived can also be successfully applied to equipment design and used in the observation and control of the growth process, at least so far as steady-state growth is concerned.

Thermal design is rather more intuitive at the present time, as a three-dimensional thermal balance calculation has not yet been performed for silicon EFG. Nevertheless, a significant amount of empirical understanding of the growth process has been generated, based on a limited one-dimensional heat transfer model for the inferface region and a simple two-dimensional model for the solid ribbon. This understanding is now being applied successfully to guide the directions in machine design and growth process optimization for the growth of wide ribbon at high rates with sufficient success that multiple growth of five ribbons, 5 cm wide at rates of 3 cm to 3.5 cm/min, has been sustained for many hours. Also, growth of 10 cm wide ribbon from single cartridge stations at similar speeds is now a reality at Mobil Tyco, and 12.5 cm wide ribbons have been grown at lower growth speeds from induction-heated equipment elsewhere[61]. It seems fairly clear that, at the time this paper appears in print, multiple growth of 10 cm wide ribbon will have been achieved.

One may also conclude at this time that the multiple cartridge concept for EFG appears to be practical for the production of silicon ribbon at least 10 cm in width at growth speeds up to perhaps 5 cm/min, while operating continuously over long periods of time. However, to summarize the experience regarding the "quality" of the ribbon materials as a basis for solar cell production is rather more difficult.

The results on cell efficiency are still somewhat confusing. Top efficiencies on small cells, prepared from material with representative distribution of the "fundamental defects" found in ribbon, have reached 14% (AM1) when using conservative cell processing schemes. Such material was prepared at speeds around 2 cm/min and 2.5 cm wide from simple induction-heated systems using quartz crucibles.

Indeed, average efficiencies of large batches of cells 2.5 cm x 10 cm in area, prepared from long single ribbons grown in this mode, have often averaged more than 11% (AM1) and, in three-shift pilot production, such machines have achieved annual averages near 10% (AM1). The material from resistance-heated cartridge type equipment, however, has only recently produced cells around 12% efficiency in small numbers. Also, this equipment generally employs carbon crucibles to hold the silicon and a different diffusion process was more successful here for cell preparation.

In general, the results on electrical properties of the ribbon at present suggest that there is still a fundamental lack in understanding such growth process variables as interface shape optimization, microsegregation effects in the meniscus, impurity complexing, and the effect of carbon (either contributed from the die or from the crucible). Thus, which of the machine design variables have significant effects on the electrical properties can only be determined by systematic empirical experimentation along with a detailed characterization of ribbon grown under various conditions. Such efforts, which have been underway for several years, are beginning now to yield more basic results which, in the near future, should make average efficiencies of large area cells (~ 50 cm^2) in the region of 13% (AM1) possible.

Such efficiencies, along with expected further improvements in ribbon throughput and automation, should therefore make EFG a prime candidate for the low-cost production of solar cell blanks for terrestrial use.

Acknowledgements. The author is deeply indebted to many colleagues at Mobil Tyco whose thoughts, concepts, and, on occasion, very words he used to write this paper. Thanks are also due to Professor Bruce Chalmers of Harvard University and Dr. Juris Kalejs of Mobil Tyco Solar Energy Corporation for critical readings of the manuscript, from which many improvements resulted.

Also, the author is grateful to Dr. D. Ast and Dr. H. J. Strunk of Cornell University for the permission to quote unpublished results and for supplying Fig. 26.

Some of the developments referred to here, in particular those relating to multiple ribbon growth, were made with partial financial support from the US. Department of Energy through the "Low Cost Solar Array Project", which is administered by the Jet Propulsion Laboratory at the California Institute of Technology.

7 References

1. Ravi, K. V., Mlavsky, A. I: The Silicon Ribbon Solar Cell – A Way to Harness Solar Energy. In: Heliotechnique et Developement Kettani, M. H., Soussou, J., (eds.), pp. 563–578, Cambridge, Mass., Mass. Development Associates 1976
2. Ciszek, T. F.: Mat. Res. Bull. *7,* 731 (1972)
3. Stormont, R., Wald, F. V.: Silicon Ribbon, Final Rep, Contract No. N00163–71–6–0711, Dec. 1971; Avail.: U S. Nat., Techn. Information Service, Springfield, VA 22151
4. Bates, H. E., Cocks, F. H., Mlavsky, A. I.: The Edge Defined Film Fed Growth (EFG) of Silicon Single Crystal Ribbon for Solar Cell Applications. In: Conf. Record 9th IEEE Photovoltaic Specialists Conf., pp. 386–388. New York, IEEE 1972
5. LaBelle, H. E., Mlavsky, A. I.: Nature *216,* 574 (1967)
6. LaBelle, H. E., Mlavsky, A. I., Chalmers, B.: Mat. Res. Bull. *6,* 571 (1971); *6,* 581 (1971); *6,* 681 (1971)
7. Hurley, G. F., Pollock, J. T. A.: Met. Trans. *3,* 397 (1972)
8. Surek, T., Chalmers, B., Mlavsky, A. I.: J. Cryst. Growth *42,* 453 (1977)
9. Bates, H. E., Jewett, D.N., Mlavsky, A. I., White, V. E.: Thick Film Silicon Growth Techniques. Final Rep.,Subcontract No. 953365, Jet Propulsion Laboratory, Pasadena, CA, April 1975. Available from: US. Nat. Techn. Information Service, Springfield, VA, 22151
10. Swartz, J. C., Surek, T., Chalmers, B.: J. Electron Mat. *4,* 255 (1975)
11. Surek, T.: J. Appl. Phys. *47,* 4384 (1976)
12. Surek, T., Rao, C.V.H., Swartz, J. C., Garone, L.: J. Electrochem. Soc. *124,* 112 (1977)
13. Surek, T., Chalmers, B.: J. Cryst. Growth *29,* 1 (1975)
14. Ravi, K. V.: J. Crystal Growth *39,* 1 (1977)
15. Ciszek, T. F., Schwuttke, G. H.: US. Patent No. 4075055 (1978)
16. Mlavsky, A. I., Serreze, H. B., Stormont, R. W., Taylor, A. S.: The Tubular Silicon Solar Cell – A New Concept for Photovoltaic Power Generation. In: Twelfth IEEE Photovoltaic Specialists Conference. pp. 160–167. New York, IEEE 1976
17. Eriss, L., Stormont, R. W., Surek, T., Taylor, A. S.: J. Crystal Growth, Special Issue on Shaped Growth, to be published 1980
18. Chalmers, B., Mlavsky, A. I., Surek, T., Swartz, J. C., Bell, R. O., Jewett, D. N., Yates, D. A., Ravi, K. V., Wald, F.: Continuous Silicon Solar Cells. Final Rep., Grant No. NSF/RANN/SE/GI-37067X/FR/75/1. Aug. 1975. Available from: US. Nat. Techn. Information Service, Springfield, VA 22151
19. Sachs, E. M.: J. Crystal Growth, Special Issue on Shaped Growth, to be published 1980
20. Kalejs, J. P., Mackintosh, B. H., Surek, T.: J. Crystal Growth, Special Issue on Shaped Growth, to be published 1980
21. Yates, D. A.: Proc. Soc. Photo Optical Instrumentation Engineers, *116,* 82 (1977)
22. Rao, C. V. H., Surek, T., Mackintosh, B., Ravi, K. V., Wald, F. V.: Large Area Silicon Sheet by EFG. In: Sun, Mankind's Future Source of Energy. (de Winter, F., Cox, M., eds.), pp 643–650, New York, Pergamon Press, 1978
23. Mackintosh, B. H., Jewett, D. N.: US. Pat. 4118197 (1978)
24. Mackintosh, B. H., Surek, T., Kalejs, J. P., Sachs, E. M., Nagy, S., Wald, F. V.: Multiple Silicon Ribbon Growth by EFG. In: 13th IEEE Photovoltaic Specialists Conf., pp. 350–357, New York, IEEE 1978
25. Ravi, K. V., Wald, F. V., Gonsiorawski, R., Rao, C. V. H., Garone, L. C., Ho, J. C., Bell, R. O.: An Analysis of Factors Influencing the Efficiency of EFG Solar Cells. In: 12th IEEE Photovoltaic Specialists Conf., pp. 182–190, New York, IEEE 1976
26. Schwuttke, G. H., Yang, K., Ciszek, T. F.: J. Crystal Growth *43,* 329 (1978)
27. Rao, C. V. H., Bates, H. E., Ravi, K. V.: J. Appl. Phys. *47,* 2614 (1976)
28. Garone, L. C., Rao, C. V. H., Morrison, A. D., Surek, T., Ravi, K. V.: Appl. Phys. Lett. *29,* 511 (1976)
29. Ravi, K. V., Serreze, H. B., Bates, H. E., Jewett, D. N., Morrison, A. D., Ho, J. C.: EFG Silicon Ribbon Solar Cells. In: 11th IEEE Photovoltaic Specialists Conf., pp. 280–289., New York, IEEE, 1975

30. Ast, D., Strunk, H. J., Cornell Univ., Ithaca, NY, Private Communication
31. Leipold, M. H. deAngelis, R. J.: Structure Development in Silicon Sheet by Shaped Crystallization. In: Photovoltaic Solar Energy Conf. 1977, pp. 872–881. Dodrecht, Holland, D. Reidel Pub. Co., 1978
32. Rao, C. V. H., Cretella, M. C., Wald, F. V., Ravi, K. V.: J. Crystal Growth, Special Issue on Shaped Growth, to be published 1980
33. Cretella, M. C., Kalejs, J. P., Mobil Tyco Solar Energy Corp., Waltham, MA: Private Communication
34. Wald, F. V., Mobil Tyco Solar Energy Corp.: Unpublished
35. Voltmer, F. W., Padovani, F. A.: The Carbon Silicon Phase Diagram for Dilute Carbon Concentration and the Influence of Carbon on Czochralski Growth of Silicon. In: Semiconductor Silicon 1973 (Huff, H. R., Burgess, R. R., eds.), pp. 75–82, Princeton, NJ, Electrochemical Soc., 1973
36. Endo, Y., Yatsurugi, Y., Terai, Y., Nozaki, T.: J. Electrochem. Soc. *126,* 1482 (1979)
37. Hanoka, J. I.: Solar Cells *1,* 123 (1979/80)
38. Ast, D., Cornell Univ., Ithaca, NY: Private Communication
39. Rao, C. V. H., Taylor, A. S., Mobil Tyco Solar Energy Corp., Waltham, MA: Private Communication
40. Hanoka, J. I., Bathey, B.: A combined Quantitative EBIC and Ion Microprobe Analysis of SiC Particles in EFG Ribbon. In: 14th IEEE Photovoltaic Specialists Conf., pp. 478–483 New York, IEEE 1980
41. Nozaki, T., Yatsurugi, Y., Akiyama, N., Endo, Y., Makida, Y.: J. Radioanalytical Chem. *19,* 109 (1974)
42. Chalmers, B., Harvard University, Cambridge, MA: Private Communication
43. Kalejs, J. P.: Displaced EFG Die, US. Patent Pending
44. Ast, D., Cornell University , Ithaca, NY: Private Communication
45. Kalejs, J. P., Freedman, G. M., Wald, F. V.: J. Crystal Growth *48,* 74 (1980)
46. Kalejs, J. P.: J. Crystal Growth *44,* 329 (1978)
47. Schwuttke, G. H., I.B.M. Corp., Fishkill, NY: Private Communication
48. Sachs, E. M., Surek, T.: J. Crystal Growth, Special Issue on Shaped Growth, to be published 1980
49. Sachs, E. M.: J. Crystal Growth *47,* 477 (1979)
50. Mackintosh, B. H., Morrison, A. D., Rao, C. V. H., Ravi, K. V., Serreze, H. B., Surek, T., Wald, F. V., Yates, D. A.: Large Area Silicon Sheet by EFG, First Quart. Rep., 1977 on Subcontract No. 954355 Jet Propulsion Laboratory, Pasadena, CA, March 1977. Available from: US. Nat. Techn. Information Service, Springfield, VA 22151
51. Sood, A. K., Freedman, G. M., Bell, R. O., Wald, F. V.: Correlation of Minority Carrier Diffusion Length with Solar Cell Properties in EFG Silicon Ribbon. In: Solar Energy, (Lesk, A., Berkowitz, J., eds.), pp. 227–236, Princeton, NJ, Electrochemical Soc. 1976
52. Ho, C. T., Bell, R. O., Wald, F. V.: Appl. Phys. Lett. *31,* 463 (1977)
53. Rao, C. V. H., Bell, R. O., Cretella, M. C., Ho, J. C., Wald, F. V., Ravi, K. V.: Response of Defects to Illumination in Silicon Solar Cells. In: 13th IEEE Photovoltaic Specialists Conf., pp. 1213–1219, New York, IEEE 1978
54. Bell, R. O., Ho, C. T., Ravi, K. V., Wald, F. V., Muller, J. C., Siffert, P.: Ion Implanted Solar Cells from EFG Silicon Ribbons. In: 2nd E. C. Photovoltaic Solar Energy Conf., 1979 (Van Overstraeten, R., Palz, W., eds.), pp. 153–161, Dodrecht, Holland, D. Reidel Pub. Co. 1979
55. Yates, D. A., Bell, R. O.: Defect Analysis of Ribbon Solar Cells Using a Laser Scanner. In: 14th IEEE Photovoltaic Specialists Conf., pp. 1402–1403, New York, IEEE 1980
56. Gonsiorawski, R. C., Ravi, K. V.: The Influence of P-N Junction Formation Condidions on the Efficiency of EFG Ribbon Solar Cells. In: Extended Abstracts Book, Electrochemical Soc, Meeting, St. Louis, Missouri, May 1980, pp. 777–778, Princeton, NJ, Electrochemical Soc. 1980
57. Kalejs, J. P., Mackintosh, B. H., Sachs, E. M., Wald, F. V.: Progress in the Growth of Wide Silicon Ribbons by the EFG Method at High Speed Using Multiple Growth Stations. In: 14th IEEE Photovoltaic Specialists Conf., pp. 13–18, New York, IEEE 1980

58. Gonsiorawski, R., US. Patent No. 4152824 (1979)
59. Chamberlain, R. G.;: Recent Developments in the Economic Modeling of Photovoltaic Module Manufacturing. In: 2nd E. C. Photovoltaic Solar Energy Conf. 1979 (Van Overstraeten, R., Palz, W., eds.), pp. 851–858, Dordrecht, Holland, D. Reidel Pub. Co. 1979
60. Morse, F. H.: Photovoltaics Commercialization Readiness Assessment. In: 2nd E. C. Photovoltaic Solar Energy Conference 1979. (Van Overstraeten, R., Palz, W., eds), pp. 843–850, Dordrecht, Holland, D. Reidel Pub. Co. 1979
61. Gessert, C., and Schwuttke, G. H.: Wide Silicon Ribbon Growth by Capillary Action Shaping Technique. In: 2nd E. C. Photovoltaic Solar Energy Conf. 1979 (Van Overstraeten, R., Palz, W., eds.), pp. 123–129, Dordrecht, Holland, D. Reidel Pub. Co. 1979

Author Index Volumes 1–5

The volume numbers are printed in italics

Crystals

Growth, Properties, and Applications
Managing Editor: H. C. Freyhardt

Volume 1
Crystals for Magnetic Applications
Editor: C. J. M. Rooijmans
1978. 79 figures, 8 tables. VI, 139 pages
ISBN 3-540-09002-9

Contents:
W. Tolksdorf, F. Welz: Crystal Growth of Magnetic Garnets from High-Temperature Solutions (120 ref.)
F. J. Bruni: Gadolinium Gallium Garnet (49 ref.)
M. H. Randles: Liquid Phase Epitaxial Growth of Magnetic Garnet (103 ref.)
L. N. Demianets: Hydrothermal Crystallization of Magnetic Oxides (74 ref.)
M. Sugimoto: Magnetic Spinel Single Crystals by Bridgman Technique (24 ref.)

Volume 2
Growth and Properties
Editor: H. C. Freyhardt
1980. 97 figures, 23 tables. VI, 199 pages
ISBN 3-540-09697-3

Contents:
K. Nassau, J. Nassau: The Growth of Synthetic and Imitation Gems (153 ref.)
E. Schönherr: The Growth of Large Crystals from the Vapor Phase (143 ref.)
D. E. Ovsienko, G. A. Alfintsev: Crystal Growth from the Melt/Experimental Investigation of Kinetics and Morphology (76 ref.)
A. H. Morrish: Morphology and Physical Properties of Gamma Iron Oxide (146 ref.)

Volume 3
III-V Semiconductors
1980. 92 figures, 9 tables. VI, 163 pages
ISBN 3-540-09957-3

Contents:
K.-W. Benz, E. Bauser: Growth of Binary III-V Semiconductors from Metallic Solutions (149 ref.)
K. Gillessen, A. J. Marshall, J. Hesse: Temperature Gradient Solution Growth/Application to III-V Semiconductors (59 ref.)
K. Ploog: Molecular Beam Epitaxy of III-V Compounds (252 ref.)

Volume 4
Organic Crystals. Germanates. Semiconductors
1980. 133 figures, 23 tables. V, 221 pages
ISBN 3-540-10298-1

Contents:
N. Carl: High Purity Organic Molecular Crystals
L. N. Demianets, A. N. Lobachev, G. A. Emelchenko: Rare Earth Germanates
H. Maier, J. Hesse: Growth, Properties and Applications of Narrow-Gap Semiconductors

Springer-Verlag
Berlin
Heidelberg
New York

A. F. Williams

A Theoretical Approach to Inorganic Chemistry

1979. 144 figures, 17 tables. XII, 316 pages
ISBN 3-540-09073-8

Contents: Quantum Mechanics and Atomic Theory. – Simple Molecular Orbital Theory. – Structural Applications of Molecular Orbital Theory. – Electronic Spectra and Magnetic Properties of Inorganic Compounds. – Alternative Methods and Concepts. – Mechanism and Reactivity. – Descriptive Chemistry. – Physical and Spectroscopic Methods. – Appendices. – Subject Index.

This book is intended to outline the application of simple quantum mechaniscs to the study of inorganic chemistry, and to show its potential for systematizing and understanding the structure, physical properties, and reactivities of inorganic compounds. The considerable development of inorganic chemistry in recent years necessitates the establishment of a theoretical framework if the student is to acquire sound knowledge of the subject. An effort has been made to cover a wide range of subjects, and to encourage the reader to think of further extensions of the theories discussed. The importance of the critical application of theory is emphasized, and, although the book is concerned chiefly with molecular orbital theory, other approaches are discussed. The book is intended for students in the latter half of their undergraduate studies.

Springer-Verlag
Berlin
Heidelberg
New York

MIX
Papier aus verantwortungsvollen Quellen
Paper from responsible sources
FSC® C105338

If you have any concerns about our products,
you can contact us on
ProductSafety@springernature.com

In case Publisher is established outside the EU,
the EU authorized representative is:
Springer Nature Customer Service Center GmbH
Europaplatz 3, 69115 Heidelberg, Germany

Printed by Libri Plureos GmbH
in Hamburg, Germany